The Snake and the Salamander

The Snake and the Salamander

Reptiles and Amphibians from Maine to Virginia

Text by Alvin R. Breisch

Illustrations by Matt Patterson

JOHNS HOPKINS UNIVERSITY PRESS ⚮ BALTIMORE

Johns Hopkins University Press
2715 North Charles Street
Baltimore, Maryland 21218-4363
www.press.jhu.edu

A catalog record for this book is available from the
Library of Congress

ISBN 978-1-4214-2157-5 (hardback: alk paper)
ISBN 1-4214-2157-7 (hardback: alk paper)
ISBN 978-1-4214-2158-2 (ebook)
ISBN 1-4214-2158-5 (ebook)

A catalog record for this book is available from the
British Library.

*Special discounts are available for bulk purchases of this book.
For more information, please contact Special Sales at 410-516-
6936 or specialsales@press.jhu.edu.*

Johns Hopkins University Press uses environmentally
friendly book materials, including recycled text paper
that is composed of at least 30 percent post-consumer
waste, whenever possible.

To my daughters, Ariana and Kirstin,
who have been my field companions since
before they could walk

Alvin R. Breisch

To my father, David Patterson

Matt Patterson

Contents

Preface

I have had the good fortune of living for the past 20-plus years across the road from a large wetland near Albany, New York, where Sherman C. Bishop conducted detailed studies of salamanders in the 1920s and 1930s. The amphibian diversity in this area is among the highest in all of New York, with 12 species of salamanders and 8 species of frogs and toads. Spotted Salamanders, Jefferson–Blue-spotted Salamander hybrids, Spring Peepers, and American Toads breed less than 40 feet (12 m) from my door. Dr. Margaret "Meg" Stewart introduced me to the area when I was a graduate student in the late 1960s. Perhaps more than anyone, Meg awakened my interest in herpetofauna, or "herps," and showed me that it might be possible to make a career out of studying the critters I liked to play with as a kid. After I landed a job in the early 1980s with New York State as the amphibian and reptile specialist with the Endangered Species Unit, I was encouraged to contact John Behler at the Bronx Zoo and Dr. Richard "Dick" Bothner at St. Bonaventure University in Olean, New York. Both immediately became my friends and mentors, graciously offering guidance and encouragement. John taught me the fine art of turtle watching during many days in the field together. Dick was constantly explaining the right way and the wrong way to do field surveys. The take-home lesson was to put every rock or log you lift back exactly where you found it.

Since then, I have had the pleasure of working with many outstanding wildlife biologists, herpetologists, professors, and students from every state in the Northeast and many states and provinces beyond. They have all contributed to my knowledge and appreciation for this group of animals, which are often treated as second-class citizens by the general public and, in earlier years, even by the state agencies responsible for managing fish and wildlife. But I am optimistic. The plight of amphibians and reptiles has become more widely recognized over the past 30 years, and, while threats remain, attitudes toward them have definitely improved.

This book would not have been possible without the support of my wife, Sharon, who allowed me the flexibility to disappear at weird hours or sometimes for days at a time to pursue amphibians and reptiles. She also allowed my preteen daughters to accompany me on rainy, cold school nights to check which salamanders and frogs were crossing the roads near our house. Frequently accompanying me on these forays for nearly 40 years has been my friend and colleague, Mark Fitzsimmons, who more than once knocked on my door at 1:00 a.m. to let me know that it was raining and the temperature was above 40.0°F (4.4°C), often with a plastic jar containing a *Hemidactylium* or a *Pseudotriton* to entice me to leave my warm house and search the roads until dawn.

I am indebted to my friend and colleague Andy Sabin, who for more than a quarter century has provided me with guidance and support for numerous conservation projects involving amphibians and reptiles in New York, and am thankful for a generous contribution from the Andrew Sabin Family Foundation to help make this book possible. I also want to thank the Amphibian and Reptile Conservancy, Partners in Amphibian and Reptile Conservation, Edward Miller, and Don Partington for financial support to complete this book.

Thanks are also due to the staff of Johns Hopkins University Press, especially my editor, Vincent Burke, who first saw merit in producing a volume such as this, and Ashleigh McKown for numerous suggestions to improve the text, thereby enhancing its readability. Throughout the development of this book, Meagan Szekely and Kathryn Marguy provided timely responses to my questions.

Finally, I thank Thomas Pauley, whose eye for detail assured that the salamander illustrations reflected as accurately as possible the animals that Matt illustrated.

The Snake and the Salamander

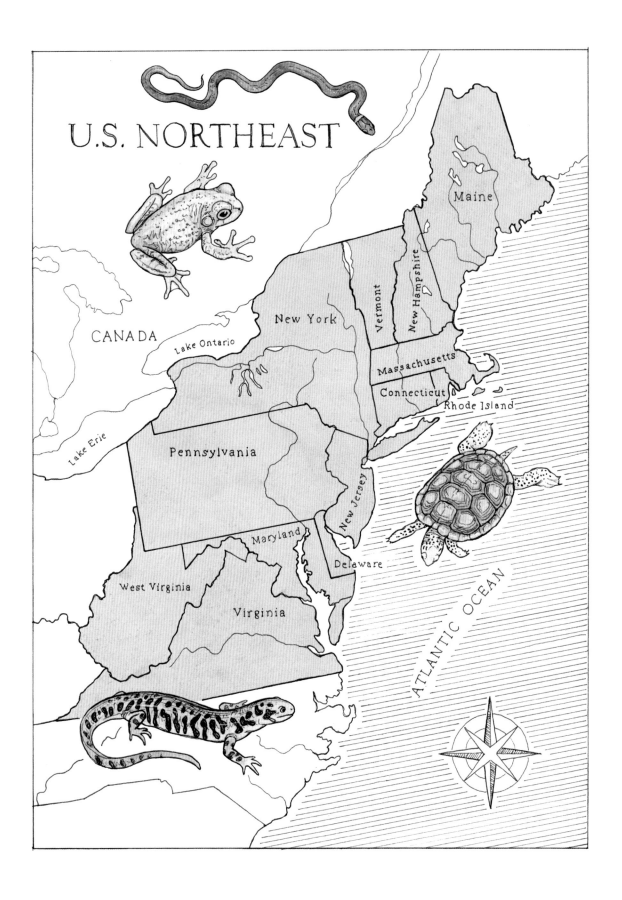

Introduction

MATT PATTERSON and I have long been fascinated by the surprising assortment of herpetofauna—the amphibians and reptiles, or simply "herps"—found here in the Northeast. Our designation of the region is the same as that adopted by both the US Fish and Wildlife Service and Partners in Amphibian and Reptile Conservation: the 13 states and the District of Columbia stretching from Maine to Virginia. This upper quadrant of America contains an amazing diversity of habitats that are the home to over 161 species of reptiles and amphibians, including slightly more than 8% of the earth's salamanders and 11% of the turtle species in the world.

When Matt first suggested we join forces to promote the beauty of these tremendous animals through his artwork and my text, I jumped at the opportunity. Rather than follow a conventional "field guide" format, we present 83 species grouped by the habitat types in which they are found. Of course, few species of wildlife use only one habitat type, so I have included a series of tables in the appendixes, cross referencing each species by the habitats they use for those who want the details. But this book is not just details; it is as much an art book as it is a guidebook on the region's herps. My short accounts that accompany each of Matt's illustrations are meant to explore some aspect of that species that I find interesting, rather than follow any structured format. I hope that Matt's original paintings will help you come to appreciate the beautiful, sometimes amazing, colors and body forms of our native amphibians and reptiles. Relatively few of these species are familiar to the general public, and some still remain mysterious and misunderstood by many naturalists and professional herpetologists.

For naturalists, the goal often seems to be to identify a species, check it off of their life list, and then move on. For some avid birders, adding a species becomes a sport. Who can see the most species in a given amount of time? For herpers, it can become the same thing. But, unlike with birds, amphibians and reptiles can often be captured, checked for key characters, and released back to the wild. With the eye of an artist, the excitement is no longer in capturing a species but in watching to see what it does when it is allowed to remain free and behave as a wild animal. I hope as you develop a greater appreciation for the beauty of these animals that instead of disturbing them, you stop and just observe, becoming a frog or a turtle or a snake watcher.

Matt's illustrations have a life-like, three-dimensional quality. They all started out as pencil drawings, with acrylic and gouache paints applied later. Most have backgrounds that are stained or toned with coffee. Matt has many reference photos for each species, showing different angles, colors, backgrounds, and behaviors. For many years, Matt has photographed all sorts of wildlife, plants, and habitats near his home in New Hampshire, so he has a large collection of his own photos to rely upon. For this project, Matt also had to obtain photos from other sources. Each of his original paintings is a composite he created.

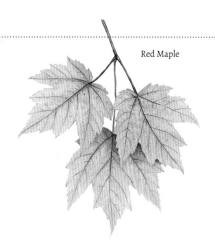

Red Maple

1 Northeastern Deciduous Forests

I STOPPED FOR a cold drink and a tank of gas at a small mom-and-pop service station just outside of Edmonton, Alberta. The owner, noticing my New York license plate, asked with a broad grin if I had ever seen mountains covered with trees before. Such misconceptions seem to be common among many people who are not familiar with the Northeast. They envision the state of New York as being just New York City, a treeless mass of tall buildings. In their minds, this treeless swath extends from Boston, Massachusetts, to Norfolk, Virginia. In truth, forests define the Northeast. The state of Maine, 90% of which is covered by woodlands, is the most forested state in the union.

The dominant landform of the region is the Ridge and Valley topography of the folded Appalachians, which extend from central Alabama through New England to the Canadian Maritimes. Of course there is also a great deal of developed land in the Northeast, ranging from isolated rural homes and small hamlets to large industrialized cities, but there are considerably more acres of forests than acres of buildings.

In colonial times, this region was nearly all forested. By the end of the nineteenth century, extensive logging and clearing for agriculture had reduced the forest to about 50% of the land area. By the 1920s, changes in the forest industry and abandonment of many farms led to a regeneration of the northeastern forest. The forest and the wildlife we see today give the impression of wild, untamed lands. But the forests and the wildlife are different. The birds, mammals, amphibians, and reptiles that relied on these forests have also suffered major declines. Fungal and insect diseases that were introduced to the area—chestnut blight, Dutch Elm disease, and hemlock wooly adelgid, to name a few—have changed the character of the forests. Gone are the American Chestnuts as a major canopy tree. Oaks, ashes, elms, hemlocks, and pines continue to decline. Gone also are the Elk, Plains Bison, and Passenger Pigeon. What looks like a healthy, mature forest is actually a recovering forest. It can never be the same.

Species whose preferred habitat is the more mature forests, such as many of the salamanders in the genus *Plethodon*, probably reached a low point in their population levels during the late nineteenth century. Loss of the forest canopy cover surrounding headwater streams would have resulted in a decline of streamside salamander populations as the waters warmed. At the same time, species that required more open habitat may have benefited. For instance, the loss of canopy would provide more nesting habitat for turtles and basking areas for snakes. Today, forests are estimated to cover 60% to 70% of the Northeast, once again changing the population dynamics of shade-tolerant versus open-canopy-loving species.

When I envision the northeastern forests, I picture near-endless landscapes of broadleaf trees mixed with Hemlock and White Pine. This is the eastern deciduous forests as described by botanist E. Lucy Braun in 1950. The northern hardwoods—American Beech, Yellow and Paper Birch, Northern Red Oak, and Sugar Maple—give way toward the south to oak-hickory-chestnut forests. With bare branches in winter, these forests are brightened in the spring by showy, flowerings trees and shrubs such as Redbud, Flowering Dogwood, Shadbush, Wild Cherry, Rhododendron, and Azalea. The flowering of Shadbush in particular corresponds to American Shad (*Alosa sapidissima*) running in the rivers and the emergence of many of the amphibian and reptile species from their winter dormancy. As trees leaf out in the spring, there is a mosaic of greens, each tree with young leaves a distinct verdant color, merging in midsummer to a rather uniform dark green. This seasonal shading of the forest floor moderates the temperature and moisture content of the soil, which benefits the amphibian species living there. It is in the fall that the hardwoods reach their full glory. Deep reds, oranges, yellows, and russet browns form a carpet covering the hills and valleys, reaching to 4,000 feet or so (1,220 m) in the mountains. These dying leaves, with the help of decomposers, will add to the soil nutrients of the ecosystem, forming the basis of the energy flow where the herps serve as both predator and prey.

Above the hardwood is a spruce-fir zone capped by alpine flora on the highest summits. The highest of these summits in the Northeast is Mount Washington, New Hampshire, the crown of the Presidential Range, at an elevation of 6,288 feet (1,917 m). Other significant alpine zones occur in the Mount Katahdin (5,270 feet, or 1,606 m) area of Maine, the Green Mountains with Mount Mansfield at 4,393 feet (1,339 m) in Vermont, and the Adirondacks with Mount Marcy at 5,343 feet (1,629 m) in New York. Combined, these alpine zones add up to a fraction of 1% of the region. Farther south there is no true alpine, but numerous grass-covered, rocky summits known as "balds" mimic the open mountaintop habitat. Spruce ridges are a feature of Virginia (Mount Rogers, elevation 5,729 feet, or 1,746 m) and West Virginia (Spruce Knob, elevation 4,863 feet, or 1,482 m). But there are no herp species unique to the spruce-fir zone. They can all be found in the mixed hardwood-conifer forests at slightly lower elevations.

Because they are so dominant in the northeastern landscape, hardwood forests provide habitat for more amphibian and reptile species than any other habitat type. Specifically, 112—or about 70%—of the 161 or so native species in the Northeast spend all or a portion of their lives in hardwood forests. Hardwood forests provide the matrix in which lie forest clearings, streams and ponds, lakes and rivers, bedrock outcrops, cliffs, talus (large fallen rocks), caves, and spring seeps. For the resident herp species, the microhabitats are more critical. Fallen logs and rocks provide essential cover. Small mammal burrows and hollow trees form retreats. A rich accumulation of humus formed from decaying leaves creates soil layers that more readily hold moisture than the dry pine-oak forests that develop on well- to excessively drained sandy soils.

The future of the northeastern deciduous forest looks promising. Natural succession is allowing abandoned agricultural land to be reclaimed as forest, although human development continues to take a share of this land. Modern standards for sustainably harvesting timber adopted by the forest industry will help ensure that the Northeast continues to have suitable habitat for the entire suite of herps that rely on these forests.

Timber Rattlesnake
Crotalus horridus

Type specimen described by Carolus Linnaeus in 1758, collected from the vicinity of New York City

Total length: at birth = 7.8 inches (19.7 cm) to adult = 74.5 inches (189.2 cm)

Endangered in Connecticut, Massachusetts, New Hampshire, New Jersey, Vermont, and Virginia (applies only to Canebrake morph)

Threatened in New York

Extirpated from Maine and Rhode Island

State reptile of West Virginia

The Timber Rattlesnake is one of the remaining symbols of wildness and one of four venomous reptile species found in the Northeast. Historically, the Timber Rattlesnake was found in all northeastern states, but it has now been extirpated from Rhode Island and Maine, Quebec and Ontario, as well as from parts of New Hampshire, Vermont, Massachusetts, Connecticut, and New York. Its populations are found in remote wooded mountain slopes where rock outcrops and talus slopes are centers of their overwintering communal dens and the summer basking sites used by gestating females.

Cryptic coloration—either as the yellow morph, black morph, or a variety of color variations in between—works to conceal their presence. This camouflage helps them to capture a meal as ambush predators or to avoid predators while lying motionless in the dry leaves on the forest floor. Although often described as aggressive, Timber Rattlesnakes respond to threats as many other snake species do: with defensive action, not aggression. Both color morphs have a black tail, which has earned them the common name of Velvet Tail Rattler in some areas. Those with a dark, brownish, or blackish crossband pattern, most obvious on the yellow morph, are known vernacularly as the Banded Rattlesnake. And in the South, the coastal form is known as the Canebrake Rattler.

For the first three centuries after the English settled Jamestown in Virginia and Henry Hudson opened what is now New York to the Dutch, the Timber Rattlesnake was feared and persecuted. Its bite was not something to ignore. It could be fatal, and early treatments were ineffective. This lethal reputation means that no other species in the Northeast gets as visceral a response as the Timber Rattlesnake. Although not described by science until the mid-eighteenth century, this species was well known to the Native Americans, who extracted the hollow fangs from dead rattlesnakes to use as lancets, and to the early European settlers who arrived on North American soils circa 1600.

Unlike the Native Americans already inhabiting North America, the European settlers viewed most of the plants and animals they encountered as either useful or dangerous. The useful included things they could eat or use to make clothing or to build shelter. The dangerous things were species that could cause harm. Timber Rattlesnakes were at the high end of the dangerous scale. Rattlesnakes could kill, and even today in some areas the snake is killed on sight, even though it is a protected species in most states. The Timber Rattlesnake's reputation made it the ideal symbol to place on the American Revolutionary War banner threatening England, "Don't tread on me."

Eastern Red-backed Salamander
Plethodon cinereus

Type specimen described by Jacob Green in 1818, collected from the
Hudson Highlands, New York

Total length: hatchling = 0.7 inches (1.8 cm) to adult = 5.0 inches (12.7 cm)

When I have asked a group I am leading or a class I am lecturing to what is the most abundant vertebrate in the northeastern forests, some respond by asking a question. Do you mean by numbers or by weight? Both, I respond in turn. Not surprisingly, many of these folks have never seen a Red-backed Salamander, and a significant number of them have never even heard of it. Yet it is a key species to the functioning of our northeastern forests.

Studies done at the Hubbard Brook Experimental Forest, New Hampshire, in the 1970s indicated that the Red-backed Salamander is the most abundant vertebrate in terms of both weight and numbers. In total, these salamanders weigh about twice that of all the woodland birds combined and are about equal to that of all the small mammals. It takes about 500 Red-backed Salamanders to equal a pound (453 g). Calculations based on surveys done in New York indicated that the estimates from New Hampshire might actually be low for the Northeast in general, with possibly as many as 14 billion individuals in New York alone, or about 14,000 tons (12,700,000 kg) of salamanders. This enormous number makes the Red-backed Salamander one of the dominant species shaping the northeastern forests. It is a primary source of energy flow through the ecosystem. It eats many small invertebrates, ranging in size from tiny Collembolans (springtails) to larger earthworms, snails, slugs, and spiders. In turn, many larger animals eat Red-backed Salamanders, everything from invertebrates such as centipedes and spiders, to shrews and voles, to Ring-necked Snakes and gartersnakes, to Robins. Nature's hot dog, you might say, everybody eats them. Turkey hunters have reported finding Red-backed Salamanders in the crops (where food is stored before it enters the stomach, where digestion begins) of the birds they harvest. That would be expected, since Wild Turkeys feed by gobbling up anything looking like food when they scratch through the leaf litter on the forest floor.

The color pattern—a red stripe down its back and tail bordered by dark brownish-black sides with a salt-and-pepper belly—makes the Red-backed Salamander an easy species to pick out. Usually. In some populations, a common variant is the Lead-backed Salamander, which lacks the red pigment on its back. On this form the back can be gray to almost black, or sometimes even silvery. In other individuals of the red-backed form, the sides can also be red, termed "erythristic," which is another way to say abnormally red in coloration. *Plethodon cinereus* comes in many additional color variations described as iridistic, albino, leucistic, amelanistic, and melanistic. Whether all these color morphs have specific adaptive significance has not yet been determined.

Whatever the color, Red-backed Salamanders, and other species in the same genus, do not adhere to the definition of amphibian learned in grade school: eggs laid in water, larvae develop in water, and then the larvae undergo metamorphosis so they can move out onto land to become a terrestrial adult. Red-backs are terrestrial for their entire lives; there is no aquatic stage. The female lays eggs under a rock or log in midsummer. The adult guards the eggs as they develop, protecting them from both predators and fungi that might attack the eggs. When the eggs hatch in late summer, mature larvae emerge with just a hint of external gills remaining. These hatchlings are tiny, with a length that is less than the diameter of a nickel, and a body that is thinner than a pencil lead.

The adults feed on numerous invertebrates in the leaf litter on the forest floor. Several studies have shown that in the absence of Red-backed Salamanders, the decomposers, such as earthworms, consume much of the leaf litter. Without the protective leaf litter, soil erosion and drying occur, potentially changing the character of the forest. So the next time you take a hike on a woodland trail, thank a Red-backed Salamander.

Northern Slimy Salamander
Plethodon glutinosus

Type specimen described by Jacob Green in 1818, collected near Princeton, New Jersey

Total length: hatchling = 0.8 inches (1.9 cm) to adult = 8.0 inches (20.3 cm)

Threatened in Connecticut

My initial response upon first encountering a Northern Slimy Salamander: *yes, this is indeed a slimy salamander*. But it wasn't. It was a sticky salamander. When threatened, the Northern Slimy Salamander exudes a white, foul-tasting secretion from its tail that easily discourages a would-be predator. Within seconds this secretion turns to glue. Handle a Slimy Salamander and your fingers stick tightly together. Worse yet, the white goo quickly turns dark brown, almost black, and does not wash off with mere soap and water. A potential predator is often left with a nasty-tasting meal of sticky goo while the salamander escapes.

Naturalists have long noted that Slimy Salamanders exhibited a great deal of variation across their extensive range. Genetic studies reported in the late 1990s demonstrated that the animal originally referred to as the Slimy Salamander was actually 13 distinct species. The most prominent of these species in the Northeast is the Northern Slimy Salamander. Two of the other species, the Atlantic Coast Slimy Salamander (*Plethodon chlorobryonis*) and the White-spotted Slimy Salamander (*P. cylindraceus*) are found in our region but are restricted to Virginia. The other 10 species are found farther south to central Florida and west to central Texas.

Slimy Salamanders emerge from hibernation several weeks or a month after the initial burst of activity by the spring-breeding amphibians: Spotted, Jefferson, and Blue-spotted Salamanders, and Wood Frogs. Because Slimy Salamanders lay their eggs on land and do not have an aquatic larval stage, you will not find them migrating across roads on their way to a breeding pond. Their migration is mostly vertical: deep into fissures or small mammal burrows to avoid winter's cold or summer's drought, to the surface for summer foraging, and just below the surface for late-summer nesting. The standard search technique for Slimy Salamanders is the same method used to find most woodland salamanders. Carefully lift cover objects to see what is hiding under them. Then return the object to the exact same position so as not to disturb the retreat or nesting sites. More often than not, I find these salamanders under rocks rather than logs. When I find a population of Slimy Salamanders, I find them in discrete pockets rather than distributed widely across the landscape. Where Slimy Salamander populations are densest, I find relatively few of the closely related Red-backed Salamanders. On a good day, when the ground is still damp from rain, I can often find several dozen in an hour. If you are lucky, you may also find the Ring-necked Snake, a common *Plethodon* predator.

Finding *P. glutinosus* by lifting rocks and logs is not the same as observing them behaving naturally in the wild. For this you need a bright light and a damp night. A warm, gentle rain is best. On these nights you will find them on the surface or partially hidden in a rock crevice, sitting in ambush, waiting for prey in the form of an insect, spider, or earthworm. But don't just look down; look up, too. Slimy Salamanders are great climbers, and on rainy, windless nights I have found them sitting high on a boulder, rock outcrop, or on the end of a branch or leaf 3.0 to 5.0 feet (0.9 to 1.5 m) above the ground. In a closed-canopy Sugar Maple–Northern Red Oak forest, being eye to eye with a Slimy Salamander in the rain after dark is a near-perfect evening.

Wehrle's Salamander
Plethodon wehrlei

Type specimen described by Henry Weed Fowler and Emmett Reid Dunn in 1917, collected from Licks Hills, Indiana County, Pennsylvania

Total length: hatchling = 0.6 inches (1.6 cm) to adult = 6.6 inches (16.8 cm)

Wehrle's Salamanders are found in a band that extends from north-central North Carolina northward to the southwestern portion of New York that was unglaciated during the Wisconsinan Period, the last glacial epoch that ended about 14,000 years ago. Similar in appearance to the Slimy Salamander, this dark-bodied salamander has irregular white and bluish to gold flecking restricted to the sides of the body. Young and adult Wehrle's Salamanders in the southern portion of their range may have red, orange, or yellow dots on the otherwise plain back. The Wehrle's throat is white compared to the dark gray throat of the Slimy Salamander. *Plethodon wehrlei* also tends to be more slender than *P. glutinosus*.

If you examine the sides of most salamanders, you will notice a series of bumps separated by rib-like indentations that extend vertically from the belly to the back. The bumps are called costal ridges, and the indentations are termed costal grooves. The ridges increase skin area for water absorption, while the grooves serve as capillary conduits to permit moisture to reach the salamander's back. The number of these grooves between the front and hind legs helps in identifying a species. The Northern Slimy Salamander has just 16 costal grooves. The Wehrle's Salamander usually has 17 or 18 but may rarely have just 16. It is a good character for separating the two

species, but it is not always infallible. If the count is 17 or 18, it is a Wehrle's; if 16, then check the color of the throat and location of the flecking to distinguish Wehrle's from Slimy. And be sure to check where you are. Wehrle's has a much more restricted range than the Northern Slimy Salamander.

Wehrle's Salamanders are most likely to be found in mature forests and tend to be in slightly drier conditions than the closely related Red-backed Salamander. Like the other species in the genus *Plethodon*, Wehrle's Salamanders are an upland species and lay their eggs on land, not in the water. The larval period is essentially restricted to the time in the egg. Emergence is therefore direct development from egg to terrestrial form, producing a free-living hatchling that does not require an aquatic period to complete development. Wehrle's Salamanders are unusual for a woodland salamander in that they lay their eggs in the winter rather than in late spring or summer as most other *Plethodons* do. Laying eggs in the winter gives the female an advantage in that she can forage throughout the entire next growing season, building up her body mass until she is ready to reproduce again the following winter. Most of the other woodland salamanders are on a biennial reproductive cycle.

Yonahlossee Salamander
Plethodon yonahlossee

Type specimen described by Emmett Reid Dunn in 1916, collected near Linville, North Carolina

Total length: hatchling unknown; adult = 8.7 inches (22.1 cm)

The Yonahlossee Salamander is an impressive *Plethodon* in both size and color. It is the largest *Plethodon* and is immediately recognized by the broad, irregular chestnut-brown or reddish band that extends from its neck onto its tail. Below the band on each side is a light gray or whitish stripe, with a belly that is darker gray. Its head and posterior two-thirds of the tail are black or with a few light flecks. Of all northeastern herps, this is one of the most poorly known. No one has reported finding a nest or a hatchling of a Yonahlossee Salamander.

The Yonahlossee Salamander's primary range is restricted to a few mountainous areas in western North Carolina, but it does extend into southwestern Virginia and the northeasternmost tip of Tennessee. Species with such a small geographic area are generally vulnerable to the conflagrations of human development. Fortunately, in this region, there are large tracts of public lands owned by both federal and state governments along with private nature and hunting preserves that keep development out of much of the area. The steep, mountainous terrain also limits the possibility of future development. In spite of the ruggedness of the area, large tracts had been cleared for grazing livestock and hardscrabble farming in the past, but the impact of agriculture on these high-elevation habitats has declined in the last century. The forest industry could have an impact, however, because a damp, thick, mature forest cover is the salamander's preferred habitat.

Like the other plethodontid salamanders, the Yonahlossee Salamander is a lungless salamander. Gas exchange, specifically absorption of oxygen and expulsion of carbon dioxide, occurs through the semipermeable skin and the tissues lining the mouth. Both the skin and the mouth have an extensive network of blood vessels to assist with this gas exchange, which works most effectively when the skin is damp. As the skin dries, the salamanders have difficulty "breathing." Holding one of these animals in a dry hand causes a great deal of stress to the salamander and can be fatal. By the same token, insect sprays or lotions on the hands can also be absorbed through the salamander's skin or create a barrier to gas exchange, so competent field researchers do not expose their research animals to these stressors.

The size of the Yonahlossee Salamander makes it a good candidate for observing the nasolabial groove, another distinctive feature of plethodontid salamanders. These grooves run as vertical slits from the nostril to the upper lip. At the onset of the breeding season, the grooves become elongated, and a protuberance develops where the grooves meets the lip. The groove is lined with chemoreceptors, allowing the male to smell the female and to detect, and avoid, other males.

Green Salamander

Aneides aeneus

Type specimen described by Edward Drinker Cope and Alpheus Spring Packard
in 1881, collected from Nickajack Cave, Marion County, Tennessee

Total length: metamorph = 0.7 inches (1.8 cm) to adult = 5.5 inches (14.0 cm)

Endangered in Maryland

Threatened in Pennsylvania

The Green Salamander is found primarily in the Appalachians from southwestern Pennsylvania to Alabama and Mississippi. This species is widespread in West Virginia and southwest Virginia, with a disjunct population in the eastern panhandle of West Virginia. Green Salamanders exist mostly as small populations widely separated geographically.

Its very name stirs interest. Northeastern salamanders come in many colors, but green is not expected. This small, attractive species is in fact the only truly green northeastern salamander. The mottled green markings on a dark background give it the appearance of a lichen-covered rock, its typical home, and the place where we begin our search. Long, stout legs and feet with enlarged toe pads aid it in navigating its vertical habitat.

The Green Salamander is one of seven species in the genus *Aneides*, known collectively as the Climbing Salamanders, referring to their rock- and tree-climbing abilities. *A. aeneus* is the only species east of the Great Plains. It is a denizen of thin and damp—but not wet— rock crevices. The Green Salamander's body and head are flattened to fit this niche. Most active at night, a headlamp or flashlight is essential for the search, but even during the day these tools are helpful for peering deep into the fissures where one might see the salamander's eye looking back from deep within the crevice.

The lichen-like camouflage also makes the Green Salamander at home on tree trunks. But you will only find them on tree trunks that are near bedrock outcrops containing deep fissures. And you will only find them in bedrock crevices that are shaded by tall trees. They overwinter and nest in these deep bedrock crevices. The female reproduces every two years. She guards her eggs until they hatch in 2.5 to 3 months in late summer. Neonates and young juveniles remain close to the bedrock outcrops, while adults venture farther afield and are the ones most often found on trees. During the day, they may retreat under loose bark or in cavities in the tree as high as 69 feet (21 m) above the ground. At night, they may be more exposed and are seen on smooth-barked species such as American Beech (*Fagus grandifolia*) as well as on trees with deep furrowed bark. Whether on rock outcrops or in trees, Green Salamanders feed on insects, primarily beetles, ants, and mosquitoes.

Several authors have reported that Green Salamanders have experienced a population decline since the 1950s. The onset of this decline may be associated with the decline of American Chestnut (*Castanea dentata*) due to chestnut blight, which changed the character of eastern forests, killing an estimated four billion trees in the early part of the twentieth century.

Northern Coal Skink
Plestiodon anthracinus anthracinus

Type specimen described by Spencer Fullerton Baird in 1850, collected from
North Mountain, Carlisle, Pennsylvania

Total length: hatchling = 1.9 inches (4.8 cm) to adult = 7.0 inches (17.8 cm)

Endangered in Maryland

Skinks, formerly in the genus *Eumeces*, are a type of lizard with smooth, shiny scales. The combination of these characters with speed and agility in escaping makes them difficult to hold when captured, and difficult to examine closely for identification. In the wild they rarely stand still long enough for a good look. Fortunately, there are relatively few lizards in the Northeast, so the task of identifying a Northern Coal Skink is not that difficult. It is the only four-lined skink within its northeastern range, which extends as a number of widely separated populations from western New York to western Virginia and eastern West Virginia.

The Coal Skink's specific name, *anthracinus*, derives from anthracite, or hard coal, the preferred coal with the highest carbon content that produces the most heat with the fewest impurities. But the area where the Coal Skink was first reported is not in the region of Pennsylvania where that kind of coal is found. Instead, it got its name from the very dark brown to nearly coal-black sides. The dark lateral band is bordered both above and below by a thin light stripe extending well onto the tail.

During mating season, in middle to late spring, the side of the male's head and neck turn an orangish red.

After mating, the female lays six or seven eggs under logs or rocks in early summer and then guards them until they hatch about a month later, in midsummer. The hatchlings look similar to the adults, with the dark body and obvious light stripes except for the last third of the tail, which is a striking blue. The blue is a distraction or warning to predators. The warning is "I might not taste good." The distraction is that the blue color is the first thing a predator sees, and if the predator grabs the skink by the tail, it simply breaks off and the skink escapes.

The Coal Skink can be found in young, second-growth forests with open canopies to densely shaded hardwoods or mixed hardwoods. In all cases the habitat is usually not far from water or is at least located on a damp, rocky hillside. When startled, Coal Skinks readily take to water to avoid capture. Contrary to the general habitat pattern, its northernmost population, in Upstate New York, is a roughly 4,000-acre (1,620-hectare) forested wetland where the skink hides in slightly raised, mossy hummocks topped with short shrubs. When approached, they retreat either deep into the moss or into the shallow marl pools found throughout the wetland.

Northern Rough Greensnake
Opheodrys aestivus aestivus

Type specimen described by Carolus Linnaeus in 1766, collected near Charleston, South Carolina

Total length: hatchling = 6.8 inches (17.2 cm) to adult = 45.6 inches (115.9 cm)

Endangered in Pennsylvania

One of two unpatterned green snakes found in the Northeast, the Rough Greensnake has keeled, rather than smooth, scales and is by far a better climber than the Smooth Greensnake. The Rough Greensnake is also a more southern species, barely reaching as far north as a few populations on the southern border of Pennsylvania and the southern half of New Jersey. It is, for the most part, absent in the higher elevations of the Appalachians. The distribution of the Smooth Greensnake is almost the opposite, found in higher elevations and from southern Pennsylvania north, with little overlap in the range of the two species. As with the Smooth Greensnake, the color of a Rough Greensnake fades to a light blue within hours after death.

An excellent climber, the Rough Greensnake can be found climbing or basking on vines, shrubs, and small trees, especially those along or overhanging watercourses. From its shoreline perch, it readily takes to water. At night, it sleeps perched on branches in a coil. A typical search technique is to use a spotlight or flashlight while wading or slowly drifting in a boat along the shore at night and shine it toward the branches of the woody vegetation lining the pond or stream. In daylight, the green coloration makes it hard for the human eye to detect a Rough Greensnake either on the ground in grass or while it is lying motionless in a leafy tree. If startled or handled, it assumes a defensive posture, opening its mouth in a wide gape to display the dark purplish-black lining.

Rough Greensnakes exclusively eat invertebrates, with their primary foods being spiders and insect such as crickets, grasshoppers, and the larvae of moths and butterflies. They actively hunt prey while slowly gliding along branches of woody vegetation. Several authors have suggested that the use of pesticides that kill prey species has contributed to a decline of this species over the last few decades. Feral cats and domestic cats left loose to run are also considered a threat to both Rough and Smooth Greensnakes.

Rough Greensnakes mate in both the spring and the fall, with the female laying 2 to 14 eggs in June or July. Eggs are usually laid inside decaying logs or under rocks and boards in soft soil. They have also been reported nesting in hollow trees as much as 10 feet (3 m) above ground. A female Rough Greensnake may return to nest in the same cavity in a tree for several years.

Ring-necked Snake
Diadophis punctatus

Type specimen described by Carolus Linnaeus in 1766, collected from Charleston, South Carolina

Total length: hatchling = 4.0 inches (10.2 cm) to adult = 27.7 inches (70.4 cm)

The Ring-necked Snake is the most slender snake in the region. The plain dark gray, almost black, back with smooth scales gives it a satiny appearance. A bright yellow ring around its neck joins with a bright yellow belly. The ventral surface of the tail is often reddish. A few small black dots can be seen in the center of the belly scales of the northern subspecies, *Diadophis punctatus edwardsii*. The southern subspecies, *D. punctatus punctatus*—found in southern New Jersey, the Delmarva Peninsula, and southern Virginia and south—has a central row of crescent-shaped spots on its belly scales and an incomplete ring that doesn't quite join at the base of the head. The southern subspecies is smaller than its northern cousin, reaching a maximum length of only 18.9 inches (48.0 cm). When handled, a Ring-necked Snake assumes a defensive posture, curling its tail into the shape of a spiral curly fry while exposing its brightly colored belly. In addition to two subspecies, there are ten other subspecies of Ring-necked Snake in North America.

I often find members of this woodland species on rocky, second-growth forested hillsides, road cuts, or old shale borrow pits. I also expect that where I find Ring-necked Snakes, small *Plethodon* salamanders, one of their favorite foods, will not be far away. They also prey on earthworms, slugs, and small lizards. Ring-necked

Snakes use constricting as a way to hold and kill prey. Unlike humans, who replace their baby teeth just once to grow adult teeth, Ring-necked Snakes can replace their teeth many times during their life. Most snakes have four rows of teeth on their upper jaw and two rows on their lower jaw. Their teeth point backward on their jaws toward their throat. This geometry makes it difficult for prey to escape from the snake's grasp once it is securely grasped. In the struggle with the prey, however, these delicate teeth break off easily and are swallowed. New teeth quickly replace the lost ones.

The Southern Ring-necked Snake and all other subspecies, except *D. punctatus edwardsii*, are rear-fanged snakes that use envenomation in combination with constriction to secure their prey. The southern subspecies is only slightly venomous. Their nonaggressive nature and small, rear-pointing fangs pose little threat to humans who handle them. The rear fangs, which are the last teeth on the upper jaw, are longer than the other maxillary teeth and are connected to the Duvernoy's gland by a groove. Duvernoy's gland is in some ways analogous to the venom gland in pit vipers. But the secretions from this gland aid in swallowing and digestion rather than killing the prey.

Northern Red-bellied Snake
Storeria occipitomaculata occipitomaculata

Type specimen described by David Humphreys Storer in 1839, collected from Amherst, Massachusetts

Total length: at birth = 2.4 inches (6.1 cm) to adult = 16.2 inches (41.1 cm)

This small, slightly stocky snake with keeled scales has a plain red belly that is usually a sufficient field character to confirm its identification. Some Northern Red-bellied Snakes lack the trademark red belly and instead have bellies that are more orange or yellowish, but three pale spots just behind the head, which at times fuse into a band crossing the top of the neck, will solidify the identification. The general body color varies from gray to a tan or dark brown to an olive purple. Two faint lines, slightly darker than the body color, separate the back from the sides, starting just behind the head and running to the base of the tail. Two more faint lines separate the sides from the belly. The sides of the snake are often a slightly darker shade than the dorsum, or upper side of the body.

The young, born live between late July and early September, are indeed small, nearly the diameter of a pencil lead with a very dark, nearly black, body color. The neck spots on hatchlings appear as a ring on first glance, making one wonder if they may be young Ring-necked Snakes. But the belly color gives them away.

Although most frequently associated with deciduous and mixed coniferous forests, Red-bellied Snakes may also be found in grasslands, meadows, swamps, and sphagnum bogs. Like most species, their habitat requirements are not specific to one type, and snakes will move through and between the various units during their annual activity cycle. The common link is prey availability and cover objects under which snakes can retreat. They spend their winters under embedded logs, in rock crevasses, or in burrows in loose soil. They are also known to hibernate in ant mounds, a potentially hazardous location for a small snake. Yet they do not enter the ant mound until the ants have also retreated for the winter, into the deepest levels of the mound. Northern Red-bellied Snakes use the levels above the wintering ants for their retreat.

Red-bellied Snakes feed on worms and soft-bodied invertebrates such as caterpillars, slugs, and snails. They are a welcome addition to gardens, where they can hide under mulch and feed on the insect, snail and slug pests that plague many gardeners. Providing protective cover around the garden in the form of scrap pieces of plywood, flat rocks, or other objects is a more organic alternative than using pesticides. And while Red-bellied Snakes may terrorize garden pests, they are completely harmless to humans.

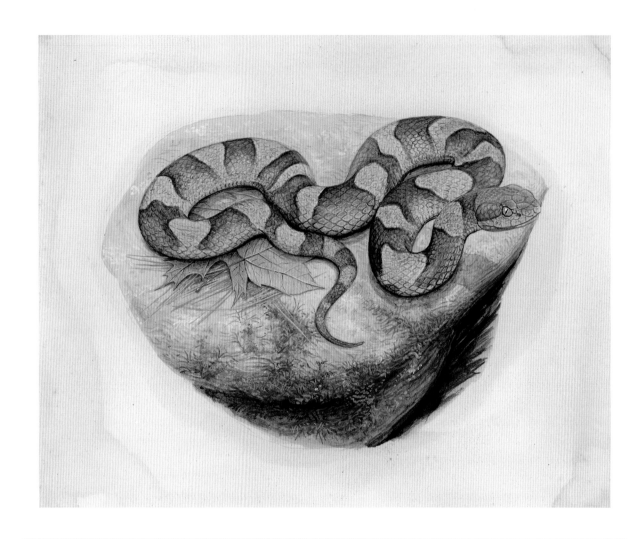

Northern Copperhead
Agkistrodon contortrix mokasen

Type specimen described by Ambroise-Marie-François-Joseph Palisot de Beauvois
in 1799, collected from the vicinity of Philadelphia, Pennsylvania

Total length: at birth = 8.0 inches (20.3 cm) to adult = 53.0 inches (134.6 cm)

The Northern Copperhead is a heavy-bodied venomous snake with weakly keeled scales, recognized by the coppery-red head that gives the snake its name. The color and patterns on its body are also quite attractive. It is marked with crossbands shaped like two Erlenmeyer flasks (like those you might find in a chemistry lab) joined at their narrow end near the midline on the back. The "flasks" are outlined with a dark chestnut-brown border and filled with a lighter chestnut color. These crossbands are separated by a lighter chestnut brown. The southern subspecies is similarly patterned but generally has a lighter shade of chestnut throughout. The intensity of the colors varies not only among individuals but also with shedding condition. Immediately after shedding, a snake's colors are at their brightest. An intergrade between the Southern Copperhead (*Agkistrodon contortrix contortrix*) and the Northern Copperhead is found in much of southeastern Virginia and the southern half of the Delmarva Peninsula.

Although its bite is serious and anyone bitten should seek medical attention, the Copperhead bite is not as dangerous as that of a Timber Rattlesnake. The Copperhead does, however, have a reputation of being something to fear, and as a result many nonvenomous snakes with crossbanded or spotted patterns on their backs have been mistaken as Copperheads and killed. This happens even to snakes that are well out of the range of where Copperheads are found. It seems that many people have the same simplistic response as my own grandfather had: "If it has a rattle, it is a rattlesnake; if it doesn't have a rattle, it is a copperhead!"

Copperheads eat mainly small rodents, as do many of the snakes that are mistaken for them. But the terms "harmful" and "harmless" as they relate to snakes fall apart here. From the point of view of a mouse or vole, they are most definitely harmful. But from the point of view of a human who recognizes that small rodents carry more diseases dangerous to man than the snakes, Copperheads and other rodent-eating snakes are all actually quite beneficial. The fact that Copperheads and other snakes eat mice, which can carry ticks infected with Lyme disease or other diseases that affect humans, should be enough incentive to allow the snakes to coexist.

The cryptic coloration of the Copperhead allows it to lie motionless and fully exposed in dry leaves, next to a log, or under a small shelter rock and not be noticed by passing predators, prey, or humans. Whereas an adult Copperhead may lie in the ambush posture next to a log with its chin touching the log, waiting for prey to pass by rather than actively seeking them, the newborn Copperheads have a different strategy. They are born with bright yellow tail tips, which they wave sinuously like a worm to lure small prey.

2 Dry Pine Woodlands

D RY PINE woodlands thrive under xeric conditions, that is, where little moisture is available. These open-canopied forests generally develop on soils that are often excessively well drained, acidic, and nutrient poor. There is insufficient soil moisture available for most broadleaf trees to thrive. This ecological community is called a "barrens" by most authorities, and it generally forms on areas where the soil is sandy, glacial deposits, or in areas of thin soils where bedrock is at or very near the surface. In spite of the harsh nature of these xeric habitats, 108 species of amphibians and reptiles frequent this habitat type in the Northeast for a significant portion of their annual activity cycle. The primary tree species occupying barrens are Pitch Pine (*Pinus rigida*), which is found throughout most of the region; Jack Pine (*Pinus banksiana*) and Red Pine (*Pinus resinosa*), occurring in the more northern barrens; with Loblolly Pine (*Pinus taeda*) and Virginia Pine (*Pinus virginiana*) in the southern barrens. Tall deciduous trees are mostly lacking, but in southern barrens, Blackjack Oak (*Quercus marilandica*) is often present. An understory of scattered, drought-resistant shrubs such as the scrub oaks (*Quercus ilicifolia* and *Quercus prinoides*), Sweetfern (*Comptonia peregrina*), and lowbush blueberries (*Vaccinium* species), as well as prairie grasses and wildflowers, may be present.

Pine barrens are globally rare ecosystems supporting a number of rare plants and animals in the Northeast. Some of the more uncommon and rarely seen amphibians and reptiles are adapted to live in barrens ecosystems, where the pine-oak savannah is interspersed with shallow seasonal pools, small streams, rivulets, and spring seeps. The most distinctive pine barrens amphibians are the Spadefoot and Fowler's Toads and the Pine Barrens and Green Treefrogs. Pine barrens reptiles include the Northern Pinesnake (*Pituophis melanoleucus*) and the Northern Scarletsnake (*Cemophora coccinea copei*). Wetlands scattered throughout the barrens serve as breeding habitat for the amphibians that spend most of the year in the surrounding forests. These xeric pine forests, with abundant rotten stumps and downed logs, offer summer foraging habitat and overwintering retreat sites for many species. Some species, such as the Eastern Hog-nosed Snake (*Heterodon platirhinos*), are adapted to these well-drained sandy habitats in which they excavate their own burrows for nesting and overwintering.

Periodic fires are an essential component of the barrens ecosystem. Naturally occurring fires help maintain the vegetation in an early to midsuccessional stage. The well-spaced pines are relatively thick barked and fire resistant, with canopies that do not touch adjacent trees, limiting the ability of fire to spread from treetop to treetop. In the absence of fire, an open understory of oak scrub develops, and, if fire is suppressed for decades, oak trees or more mesic hardwoods eventually dominate the forest. Where the natural fire regime is suppressed, periodic prescribed burns are necessary to maintain this interesting assemblage of flora and fauna.

The best known of the pine barrens areas in the Northeast are the 1.1 million acres (4,500 km²) of the Pinelands National Reserve in New Jersey and the 100,000-acre (405 km²) Long Island Central Pine Barrens in New York, both of which are dominated by Pitch Pine and serve to protect immense drinking-water aquifers.

Fowler's Toad
Anaxyrus fowleri

Type specimen described by Mary Hewes Hinckley in 1882, collected near
Milton, Massachusetts

Head–body length: metamorph = 0.3 inches (0.7 cm) to adult = 3.7 inches (9.5 cm)

The Fowler's Toad is somewhat smaller and has a more restricted range than its cousin the American Toad. It is absent from northern Pennsylvania, northern New York, and northern New England. It is the common toad along the Atlantic coastal plain and the only true toad found on Long Island, southern New Jersey, and much of the Delmarva Peninsula. It is found almost exclusively in areas with sandy soil, which restricts the species distribution and abundance at interior sites. The call of the Fowler's Toad is an unmusical, nasal bleating of shorter than five seconds, sounding not unlike a herd of goats or sheep.

The Fowler's Toad's ventral surface is plain white or with a single large black spot on its chest. On its back are scattered black spots and a light-colored middorsal stripe. Within the larger black spots are three or more, rarely only two, orange or reddish-brown wart-like bumps. The general color is brown or gray, but sometimes these toads may appear to be dull reddish or even have a greenish hue.

On the top of its head behind the eye is an L-shaped ridge, the cranial crest, which separates the eye from, and touches, a peanut-shaped mound, the parotoid gland. The parotoid gland produces a moderately potent toxic, milky-white chemical, bufotoxin, that is a defense against predators. The chemical, which is present in all species of *Anaxyrus* (formerly *Bufo*) acts as a neurotoxin and cardiac stimulant. It is quite distasteful to predators. Bufotoxin may cause hallucinogenic behavior and possibly death if too much is ingested.

As toxic as the bufotoxins are, some species of predators have evolved a resistance to them. Toads, and Fowler's Toads in particular, are a favorite food of Hog-nosed Snakes. Watersnakes, gartersnakes, raccoons, and skunks also prey on these toads. The poison is toxic to humans if ingested or is an annoying irritant if it comes in contact with skin cuts or abrasions or mucous membranes, especially in the eyes. This effect may have contributed to the mythology that touching a warty toad causes warts.

Toads have short legs so are not great leapers and jumpers, but instead they move in small hops, giving them the moniker of "hop toad." Unfortunately, they must make migrations each spring from their overwintering sites in the uplands to their breeding ponds, a journey that in the Northeast often takes them across roadways. Their tendency to hop a few hops and then stand erect for a few seconds or minutes results in large numbers being killed by vehicles each year. In many instances, nighttime or early morning scavengers—such as skunks, raccoons, opossums, foxes, owls, and crows—clean the road surface of many roadkilled amphibians, so a census of the previous evening's carnage is incomplete if delayed until midmorning. But the unpleasant-tasting chemicals found in the skin glands of toads, and also newts, often means that their carcasses are not carried away, so paper-thin shells of these species can be found several days after they were flattened.

Eastern Box Turtle
Terrapene carolina

Type specimen described by Carolus Linnaeus in 1758, collected from the vicinity of Charleston, South Carolina

Carapace length: hatchling = 1.1 inches (2.7 cm) to adult = 7.8 inches (19.8 cm)

Endangered in Maine

The Box Turtle has been called the Land Turtle, or sometimes in earlier days the Land Tortoise or Box Tortoise. Here in the Northeast, it is the most terrestrial of chelonians. Its high-domed shell and lumbering gate is more reminiscent of the true tortoises of the southern United States, the Galapagos, or Madagascar. The high-domed shell sets it apart from other northeastern turtles, with the exception of the Blanding's Turtle that shares its helmet-shaped profile. The Eastern Box Turtle, however, has the more colorful carapace, with streaks or mottling of yellow to orange highlighting the center of the dark brown to blackish scutes.

The sexes are distinguished by several characters. Generally, the female has brown eyes and the male has red eyes. The female's tail is shorter than the male's, and the male's plastron is deeply concave while the female's plastron is nearly flat. In most turtles, the plastron is attached to the carapace with a specialized structure called the bridge. This rigid bony structure allows little flexibility but does permit the legs to extend when walking or swimming while allowing the legs to be pulled in when resting or the turtle feels threatened. A unique feature of the Box Turtle is its ability to close its plastron. The bridge and the front lobe of the plastron are not as rigidly fixed as in most other turtles, allowing the plastron to be closed tightly against the carapace. This makes the adult Box Turtle predator resistant, if not quite predator proof. Large predators with strong jaws like the coyote can crush a Box Turtle, but smaller ones like a raccoon can only "play" with the turtle when it becomes a sealed box. The Box Turtle survives with perhaps just a few indents in its shell from pointed canine teeth.

While retreat into and closing of its shell has served the Box Turtle for centuries—eons, actually—these features do not protect it against man. As with most herps, loss of habitat is a major threat, and roads can be an even bigger threat. In the Northeast, the Eastern Box Turtle is completely terrestrial except for short jaunts to shallow pools during the heat of the summer. It is found most often in the open grassy areas in the summer but retreats to the forest for the winter, digging just under the forest duff layer, where it spends the cold season in a hibernation-like state. In the Northeast, these jaunts often mean encountering a road. As early as 1860, Henry David Thoreau noted that Box Turtles were being killed by wagon wheels on dirt roads. More recently, in 1941, Will Cuppy, in his delightful little book *How to Become Extinct*, noted, "The Box Turtle is the kind you run over on the road. He has a hinge on his plastron, or lower surface, enabling him to shut himself completely inside his shell and be run over." If Box Turtles were having trouble crossing dirt roads during the days of horse-drawn wagons or the slower pre–World War II autos, how would we expect them to cross high-speed two-lane roads, multiple-lane interstate highways, or the eight- to ten-lane divided highways that serve as conduits into our cities?

I grew up in southeastern Pennsylvania, and the Eastern Box Turtle was my favorite turtle. It was the one we saw most often in our rural farming community, and I and other kids would often bring one home when we found one while out hiking across nearby fields or biking on nearby roads. About 20 years after I left home to attend college, my dad wrote me a letter in which he noted sadly that it had been 10 years since he had seen a Box Turtle in the area. The rural farming landscape was being lost to residential and commercial developments.

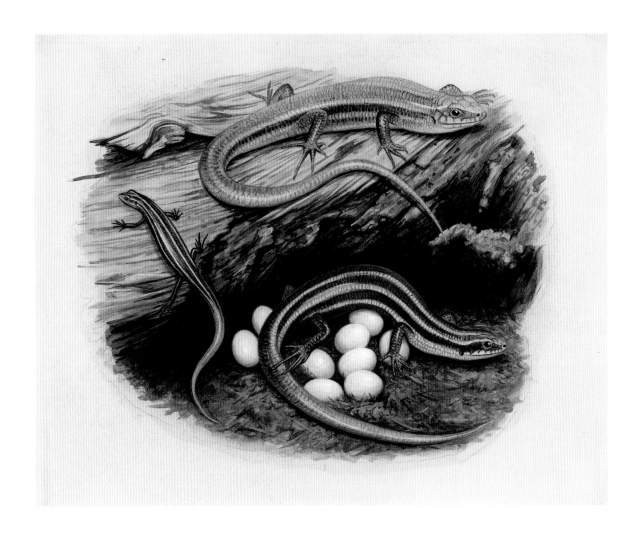

Common Five-lined Skink
Plestiodon fasciatus

Type specimen described by Carolus Linnaeus in 1758, collected from the vicinity of Charleston, South Carolina

Total length: hatchling = 2.0 inches (5.1 cm) to adult = 8.5 inches (21.5 cm)

Endangered in Vermont

Threatened in Connecticut

Extirpated from Massachusetts

If you were to encounter an adult male, an adult female, and a juvenile Five-lined Skink, you might think you had found three distinct species. The five thin, cream-colored to light yellow stripes on a dark body are most obvious on the juvenile. The stripes on the female become duller as the color of her body lightens with age, but she does retain the overall striped pattern. The light stripes are lost on the old male as they blend with the general brown shade of his back and slightly darker shade on the sides. During breeding season, the male's jaws become a distinct orange red.

The juvenile Five-lined Skink has a cobalt-blue tail, which sets him off from the adults. The common name given the skink by early European settlers, Blue-tailed Skink, applies not just to this species but also to the Coal Skink (*Plestiodon athracinus*), Southeastern Five-lined Skink (*P. inexpectatus*), and Broad-headed Skink (*P. laticeps*). The blue tail is a warning to potential predators that the young skink might not taste good, or it could serve as a distraction, because the predator focusing on the blue is left with the tail, which breaks off easily as the skink slithers away. People who have tried eating Blue-tailed Skinks reported becoming extremely nauseous after eating just one. The early Dutch settlers in New Netherlands, what we now call New York, learned that the Native Americans greatly feared the Blue-tailed Skink because "this kind will crawl up into their fundamentals, when they lay asleep on the ground in the woods, and cause them to die in great misery." There does not seem to be any logic or reality as to how such a myth could arise, or for that matter why a skink would crawl up into someone's "fundamentals" whether they are asleep on the ground or not.

The Five-lined Skink is widespread in the southern two-thirds of Pennsylvania, with an arm extending into New York up the Hudson River watershed into the southern Adirondacks. The skink then jumps across the watershed boundary into the Lake Champlain drainage, reaching its northern limit for our region. The Five-lined Skink is also found in Ontario, touching the border of New York at Niagara Falls and again in the Thousand Island area of the St. Lawrence River. It would seem logical that the skink could have entered New York from these two points, but none have yet been found on the US side of the border in these areas.

Broad-headed Skink
Plestiodon laticeps

Type specimen described by Johann Schneider in 1801, collected near Charleston, South Carolina

Total length: hatchling = 2.3 inches (5.7 cm) to adult = 12.8 inches (32.4 cm)

The Broad-headed Skink, another member of the lizard family Scincidae, is characterized as having smooth, shiny scales. It is fast moving, difficult to catch, and even harder to hold, with a tail that can be easily broken off if handled roughly. The uninitiated often confuse salamanders with skinks. Skinks are reptiles armed with sharp claws and are encased in a body protected by scales, providing protection against desiccation. Skinks, and lizards in general, have lungs for oxygen and carbon dioxide exchange with the atmosphere. Salamanders are amphibians, a group without claws or scales. Salamanders have a semipermeable membrane that permits gases and moisture to be absorbed directly through the skin. Plethodontid salamanders, which do not have lungs, conduct respiration through the skin and can only "breathe" if the skin retains sufficient moisture. Skinks, and lizards in general, can thus occupy habitats that are more xeric than those where most salamanders are found.

The Broad-headed Skink is the largest skink in the East. Its red-orange head has earned it the vernacular name "red-headed scorpion," but of course it is not a scorpion and it is not venomous. The proportionately wider head than other skinks is more pronounced on the male, which is generally larger overall than the female.

The swollen jowls and brighter colors of the male's head during breeding season exaggerate its head width even more. The orange-red color of the breeding season fades by summer, leaving a generally dull brown–colored skink. The young have five yellow or cream-colored stripes and a blue tail. The blue fades as the skink matures. The stripes also fade, more so on the male than on the female.

The Broad-headed Skink is more of a natural climber than other skinks and uses its climbing ability to escape from potential predators or human intruders. If captured, this large skink can deliver a painful bite. Males especially may be found high in trees. Although this species and the Five-lined Skink are often found in dry oak-pine habitats, the Broad-headed Skink will tolerate harsher, even more xeric habitats then the Five-lined Skink. At the onset of the spring breeding season, males are often observed in rather vigorous combat, with the jaws of both males locked onto the flanks of their opponents. After mating, the female lays her eggs in a suitable nest location in early summer. The nest could be in a rotting stump or tree cavity, under logs, in bark or leaf litter on the ground, or even inside sawdust or mulch piles. The female guards her clutch of 6 to 20 eggs until the little blue-tailed hatchlings emerge after about a month.

Eastern Fence Lizard
Sceloporus undulatus

Type specimen described by Jacob Green in 1818, collected from the vicinity of
Princeton, New Jersey

Total length: hatchling = 1.6 inches (4.1 cm) to adult = 7.2 inches (18.4 cm)

Threatened in New York

The Eastern Fence Lizard is the only rough-scaled lizard in the Northeast. The prominently keeled, overlapping scales have sharp points directed toward the posterior, giving the lizard a spiny appearance. This primarily gray to sometimes brownish species has wavy crossbands on the back and tail that are very obvious on the female, but less so or nearly absent on the male. The crossband pattern is excellent camouflage as it rests motionless on tree bark or a weathered wooden fence. The undersides and throat of the male are a bright greenish blue that appears to change in intensity with the angle of the reflected light. The northeastern form, which was previously considered a subspecies, *Sceloporus undulatus hyancinthinus*, referred to the color on the male, which observers thought was reminiscent of the color of the Hyacinth flower.

At the northern limit of the Eastern Fence Lizard's range, in southern New York, this species is found in remote, dry, open pine forests on ridgetops and is considered quite rare. Farther south, the species is common to abundant in dry, open forests and may be found scurrying around residential developments, basking on stone walls or sidewalks. But it is rarely found far from trees or tall shrubbery and is always within a short, quick dash to the cover of fallen logs, stumps, dead leaves, or other retreat sites where it can hide. Capturing a Fence Lizard requires both patience and stealth.

The Eastern Fence Lizard is an excellent climber, aided by long, sharp claws that allow a firm grip on the tree bark or fence posts it is often observed on and from which it derives its current common name. The Eastern Fence Lizard also moves with incredible speed when startled, a trait that earned it the common name of Brown Swift in the early 1800s. In the past, both skinks and Fence Lizards were erroneously believed to be venomous and were called scorpions. The Eastern Fence Lizard was known as the Brown Scorpion. Its abundant populations in southern conifer forests gave it an additional name, the Pine Lizard.

In the late 1930s and early 1940s, Carl Kauffeld, Curator of Reptiles at the Staten Island Zoo in New York City, made periodic pilgrimages to the New Jersey Pine Barrens to collect Eastern Fence Lizards to feed to some of the snakes at the zoo. At that time, Staten Island was relatively undeveloped, with habitats similar to the ones he found in the Pine Barrens. In order to save himself from having to make those long trips to southern New Jersey, Kauffeld released 29 Eastern Fence Lizards on the island, hoping to establish a colony of lizards to feed the snakes. The Fence Lizards quickly became an established, reproducing population. Today, offspring from those original 29 lizards can still be found on Staten Island, but because of numerous developments eating up the open spaces, the Fence Lizards are now confined to one state park.

Little Brown Skink
Scincella lateralis

Type specimen described by Thomas Say in 1823, collected from the banks of the
Mississippi River below Cape Girardeau, Missouri

Total length: hatchling = 1.7 inches (4.4 cm) to adult = 5.8 inches (14.7 cm)

The Little Brown Skink is aptly named. It has a plain, shiny bronze to golden or dark brown back that is bordered with a single dorsolateral stripe on each side that extends from the snout posteriorly and onto the tail. Its belly is white to yellowish. The tail on this species is proportionally longer than that of other skink species. The young look much like the adults and do not have a blue tail, as other northeastern species commonly called skinks do.

Superficially, the color pattern of the Little Brown Skink is similar to the Northern Two-lined Salamander, but that is where the resemblance ends. Reptiles and amphibians are distantly related. Salamanders, and all other amphibian groups, evolved from purely aquatic vertebrates. They retained the dependency to return to water to lay eggs and reproduce. On the other side of the evolutionary divergence were the vertebrates that developed an amniotic membrane—reptiles, mammals, birds, and dinosaurs—which severed the dependency on an aquatic environment for successful reproduction.

The Little Brown Skink was previously known as the Ground Skink, reflecting another of its characteristics: it rarely climbs. It is mostly found frenetically scurrying about in leaf litter in search of insects or other small invertebrates. It is found in a variety of vegetation types, including dry, sandy, open pine forests to mixed hardwoods and grassy areas, in which the humus and leaf litter accumulation are adequate to provide cover. A quiet observer will often hear them rustling through the detritus on the forest floor before seeing them. Rather than seeing them, all the observer might do is detect movement under the leaves and then the skink is gone.

Lizards and snakes are closely related, with both belonging to the order Squamata. The obvious physical characteristic that separates the two groups is that lizards usually have legs and snakes usually do not, but there are exceptions to both groups. Additional characteristics separating the two are that snakes lack eyelids and ear openings, whereas lizards have both. The eyelid protects the eye for species that live in a microenvironment where fine dust particles can affect vision. Of course, the individual cannot see with the eyelid closed. The Little Brown Skink has evolved a solution to that problem. It has a clear membrane, a modified scale actually, on its lower eyelid like a window, allowing the skink to see even when its eyelid is closed. Its eye is also protected from injury as it tunnels through the soil and leaf-litter debris as it searches for its next meal.

Northern Scarletsnake
Cemophora coccinea copei

Type specimen described by Georges Jan in 1863, collected from Tennessee
Total length: hatchling = 4.9 inches (12.5 cm) to adult = 32.5 inches (82.8 cm)

An attractive fossorial, smooth-scaled snake, the Northern Scarletsnake has a row of bright red saddle-shaped blotches along its back with a black border separated by yellow-white blotches. The general appearance is that of a banded snake, but the blotches do not encircle the body. Its small pointed head, typical of fossorial species, is capped with red from the tip of its snout to just behind its eyes.

The overall color pattern of the nonvenomous Scarletsnake gives the impression of a snake marked with alternating red, black, and creamy yellow bands. To humans, this pattern looks like that of the Coralsnake (*Micrurus fulvius*). But the Coralsnake has rings encircling its body, not blotches on its back. The front of the head of the Northern Scarletsnake is red, while the Coralsnake has a black head. The Coralsnake is both venomous and reportedly distasteful to potential predators, so being a mimic may afford the Scarletsnake some protection. But because the Coralsnake does not exist north of southeastern North Carolina, the protective nature of being a mimic in our area is questionable. It may in fact be a disadvantage, as humans may kill the Scarletsnake, mistaking it for a Coralsnake, even when well outside of the natural range of the species they fear. Humans have nothing to fear from this gentle snake, which will rarely bite if handled.

The Northern Scarletsnake's range is similar to that of the Cornsnake, being found in the southeastern portion of our region, with its northernmost populations in the Pinelands area of New Jersey. Its preferred habitat is the loose, well-drained soils of dry, open-canopied pine woodlands. As with many fossorial species, Scarletsnakes are more active on the surface at night. During the day, they hide in rodent burrows or take refuge under logs or other surface-cover objects.

The Scarletsnake is a constrictor, taking a variety of prey that includes small mice, snakes, and lizards. But its most unusual feeding technique is how it uses its constricting ability to prey on the eggs of other snake species, its favored food. Small snake eggs can simply be swallowed whole. But larger eggs present a challenge. The Scarletsnake has two enlarged saber-like teeth attached to the rear of the maxillary bone in the upper jaw. It uses these teeth to cut slits in the soft eggshell. Then it wraps its coils around the egg and squeezes the contents into its mouth, swallowing it with a chewing type motion.

Red Cornsnake
Pantherophis guttatus

Type specimen described by Carolus Linnaeus in 1766, collected from the vicinity of Charleston, South Carolina

Total length: hatchling = 9.0 inches (22.9 cm) to adult = 72.0 inches (182.9 cm)

Endangered in Delaware and New Jersey

The Cornsnake, previously in the genus *Elaphe*, is among the most colorful and variable snakes in the Northeast. It has a basic body color of orange, yellow, tan, or gray. A row of red blotches outlined in black down its back is the most obvious feature. Its belly has a bold checkerboard pattern, and the underside of its tail usually has two dark stripes. Cornsnakes generally adopt a fossorial lifestyle, but they are excellent climbers and strong constrictors. They feed primarily on small mammals, but they will take birds, frogs, and lizards on the ground, under the ground in rodent burrows, and in the trees. When threatened, the Cornsnake puts on a display by rising up with its head above the surface, hissing, and tail vibrating, but it seldom bites.

These characteristics of being colorful, easily handled, and accepting a variety of prey items have made the Cornsnake a popular pet. Cornsnakes adapt readily to captivity if given a terrarium with options to hide and climb. They have also been found to breed well in captivity. The ease of care, the gorgeous colors, and variability from one individual to another of wild Cornsnakes have attracted breeders to this species. As a result of selectively breeding a number of attractive and unusual color patterns, the Cornsnake is now near the top of herpetoculturists' list. Although they can reach six feet in length, Cornsnakes offer little threat to youngsters or inexperienced hobbyists, except for their ability to escape from an unsecured cage. And because healthy, beautiful, captive-bred snakes are commercially available, collection from the wild has been reduced, but of course not eliminated, even in areas where the species is protected.

Cornsnakes are currently found from the Florida Keys north to the southern New Jersey Pine Barrens area. But that is a contracted range from what it was about the time of the last ice age, the Wisconsinan, which began about 95,000 years before present (ybp), and reached its maximum southward movement about 21,750 ybp. Bones identified as belonging to a Cornsnake were found in Highland County, Virginia, dating to about 29,000 ybp. This location is well west of the closest current record for Cornsnake in that portion of Virginia. Although there are no historic records of Cornsnake from Pennsylvania, their bones were found in Bedford County, Pennsylvania, dating to 11,000 ybp. Neither of these sites was close to the glacial front, but finding their bones and other species dating to the same era helps paleontologists re-create an image of the climate and the habitats present during that period.

Eastern Hog-nosed Snake
Heterodon platirhinos

Type specimen described by Pierre André Latreille in 1802, collected from the vicinity of Philadelphia, Pennsylvania

Total length: hatchling = 4.9 inches (12.5 cm) to adult = 45.5 inches (115.6 cm)

Endangered in New Hampshire

A call came in: "What should I do? There is a cobra in my garden!" The homeowner was obviously scared and upset, but it took just a few seconds to determine she did not have a cobra invasion but rather had encountered one of the most intriguing of snakes in the region, an Eastern Hog-nosed Snake. A typical response when a Hog-nosed Snake feels threatened is to rise up, hiss, and spread the skin on its neck to appear like the textbook picture of a cobra, ready to strike with its hood expanded. Step back and the snake relaxes, drops down, and retreats to whatever cover is available. Continue to annoy it, and it goes to its second line of defense. It writhes on the ground, twisting and turning until finally rolling over on its back in a loose coil, playing dead with its mouth hanging open and its tongue hanging out. Try to flip it over, and it flips onto its back again. The act of playing opossum has presumably evolved as a great defense mechanism to keep it from being killed by potential predators. The Hog-nosed Snake is the best actor when it comes to the herp world.

At other times, approaching a Hog-nosed Snake that sees you before you see it may cause the snake to vibrate its tail, as do rattlesnakes and several other species of nonvenomous snakes. The Hog-nosed Snake lacks horny segments on its tail, so the rapid shaking does not produce a warning sound. If, however, the habitat is dry grass or leaves, the vibrating tail, contacting the vegetation, sounds much like the venomous rattler, possibly deterring any would-be predator from checking more closely. Or the sudden sound could make a casual hiker jump!

For all its bluff and posturing, pretending to be a cobra or a rattler, the stout-bodied Hog-nosed Snake is a very gentle snake. Even an average 20- to 30-inch (51- to 76-cm) snake can be hand captured without fear of being bit—unless you have just handled a toad, especially a Fowler's Toad, its favorite food. The snake may bite your fingers, but not because of aggressive or defensive behavior but rather because it is hungry and your hand smells like food. The warty skin of toads is highly toxic to many species, a great defense for the toads. But Hog-nosed Snakes have co-evolved to be tolerant of the toad toxin, bufotoxin, and feed almost exclusively on them, although they will take an occasional frog or an insect such as a cricket. Seasonally, the relationship between toads and Hog-nosed Snakes is also in sync. Recently metamorphosed toads are available at about the same time that newly hatched Hog-nosed Snakes are searching for food, and the large Hog-nosed Snake of course can feed on the larger adult toads. You might find adult Hog-nosed Snakes frequenting areas near toad breeding ponds in spring looking for a meal.

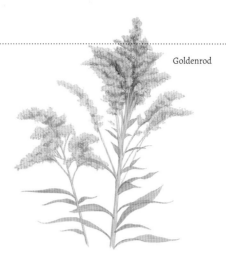
Goldenrod

3   Northeastern Grasslands

NEARLY HALF of the amphibian and reptile species in the Northeast use natural grasslands to hunt for prey, forage for berries, breed, overwinter, or simply traverse across while moving throughout their home range. In the precolonial era, grasslands in the Northeast were prairie-like landforms that extended as fingers into western Pennsylvania, New York, and West Virginia. East of those fingers were generally small, scattered areas of prairie vegetation, dominated by grasses such as Little Bluestem (*Schizachyrium scoparium*) and numerous forbs. The largest of these grasslands east of the Appalachians was the 60,000-acre (24,280-hectare) Hempstead Plain on western Long Island. The Hempstead Plain was once the home of the now-extinct Heath Hen (*Tympanuchus cupido cupido*).

The presence of Bison and Elk along with periodic wildfires maintained these open grasslands. Native Americans used fire to create early or midsuccessional vegetation-type habitat that was relatively rare in the Northeast. Beaver also played a role in clearing land around the ponds they constructed by using the fallen trees to build dams. The upland areas with canopy trees removed created open habitats for numerous herp species to carry out their life requisites. The dams also modified the local hydrology, which eventually changed siltation rates and deposition. Areas of well-drained upland grasslands and poorly drained wet meadows were formed after the beaver abandoned the area. The natural systems in the Northeast tended toward what was called the climax vegetation—that is, closed-canopy forests—the composition of which depended on altitude and latitude. There is almost no natural grassland remaining in the Northeast outside of a few high-elevation balds along the Appalachian Mountains in Virginia and West Virginia.

Rural development, clear-cutting of forests, and clearing for agriculture changed the aspect of the land. In some states, more than 60% of the land was cleared. Changes in the demand for forest products and the abandonment of agricultural land over the last century have again changed the face of the land. Today about 22% of the Northeast is now in agricultural production, including cropland, pastureland, and orchards, plus associated hedgerows, rock walls, and farm ponds, which add additional microhabitat features and retreat sites for many wildlife species.

Recently abandoned agricultural land that has not yet reverted to woodland is included in this habitat type. So long as these areas maintain an open savannah-type appearance, they function much like open grassland would. Unless some external force reverses succession, within a few decades these areas will function more as a closed-canopy forest and will no longer provide the wildlife benefits of grasslands. Unfortunately, a conflicting intention for these lands is what real estate agents term "highest and best use," or to be specific, residential and commercial developments. To anyone who enjoys the outdoors, "highest and best use" clearly has another meaning: maintaining the land as a natural community of plants and animals.

Roland M. Harper, writing in the *Bulletin of the American Geographical Society* in 1911, lamented that "no one seems to have yet attempted seriously to enumerate, classify and explain the numerous and various treeless areas of Eastern North America. . . . There are abundant hints of small prairies, open glades, natural meadows, etc., in early descriptive works dealing with parts of the country that are now pretty thickly settled, and many examples of them have doubtless already been effectually obliterated, and irrevocably lost to science."

Northern Leopard Frog
Lithobates pipiens

Type specimen described by Johann Christian Daniel von Schreber in 1782,
collected from White Plains, New York

Head–body length: metamorph = 0.7 inches (1.8 cm) to adult = 4.4 inches (11.1 cm)

Threatened in Rhode Island

State amphibian of Vermont

The Northern Leopard Frog, formerly in the genus *Rana*, was also known as the Meadow Frog, Grass Frog, or Spring Frog in earlier days. It was called the Meadow Frog or Grass Frog because it is often seen in grassy areas in midsummer. Be on the lookout for Leopard Frogs when mowing your lawn, especially when the grass is damp with dew or a recent rain. It was known as the Spring Frog because it is one of the first frogs whose choruses are heard in the spring, shortly after emerging from the pond bottom where it overwintered. The guttural snoring sound draws your attention, and then while watching the males call you notice that, unlike toads or Bullfrogs, which have a single large vocal sac under the throat, the Leopard Frog has two vocal sacs, one on each side of its throat, making it appear chipmunk-like with cheek pouches full of nuts. The Leopard Frog is the frog most frequently seen in high school biology classes. Northern Leopard Frogs have been and still are sold as bait to fishermen throughout much of the United States.

As the common name implies, this frog is spotted like a leopard, but then the Pickerel Frog is also a spotted frog. The Pickerel Frog's spots tend to be squarish rather than round, as in the Leopard Frog, and the Pickerel Frog's thighs and groin area are painted with a prominent yellow wash.

The Northern Leopard Frog is a widespread, attractive, and locally abundant frog. It stirred interest in researchers who noticed subtle differences in the frogs found in subarctic Canada to southern Florida, into Mexico, and from the Atlantic coast west to eastern California and Alberta. In the 1960s, the Northern Leopard Frog was described as a species complex composed of six subspecies, with characteristics that were not always easy to discern in the field. Careful analysis of life history traits, morphometrics, spectrograms to analyze its vocalizations, and genetic studies of the various subspecies gave further evidence that some of the described subspecies were distinct enough to be recognized as separate species. Today, a true *Lithobates pipiens*, the Northern Leopard Frog, is a species without any subspecies.

All this may mean we want to revisit the type locality for the Northern Leopard Frog. Because the exact location where von Schreber collected his specimen is not known, various authors have suggested several locations as the type locality. If the type specimen did indeed come from White Plains, a city just north of New York City, it is today outside of the range of the *L. pipiens*. Was that first frog described back in 1782 one of the other original subspecies that is now a separate species? Or did the Northern Leopard Frog have a greater range than it has today? In an attempt to clarify exactly what constitutes a Northern Leopard Frog, in 1974 Anne Pace designated a neotype—that is, a specimen selected to replace the original type specimen because it was lost—for *L. pipiens* from Etna, New York, a small town near Cornell University.

Eastern Wormsnake
Carphophis amoenus

Type specimen described by Thomas Say in 1825, collected from the vicinity of
Philadelphia, Pennsylvania

Total length: hatchling = 3.3 inches (8.3 cm) to adult = 13.3 inches (33.7 cm)

Threatened in Massachusetts

Fossorial. If asked for one word to describe the Eastern Wormsnake, it would be fossorial. Not just because it spends most of its life living underground, but more importantly because it is evolutionarily well adapted to living a subterranean lifestyle. It lacks feet to dig a burrow, but with its smooth scales and narrow, flattened, pointed head with small eyes, it can push its way underground in loose soils or tunnel into rock fissures or other animals' burrows. A specialized scale at the end of its tail is shaped like a spine, which it can use to anchor itself while pushing forward. A good general description is that the Eastern Wormsnake looks like an earthworm, a plain brown worm with an opalescent or iridescent sheen and a pink belly.

Being fossorial has a number of advantages. The temperature underground is quite constant—never too cold, never too hot—providing benefits for species that thermoregulate. As might be expected, a small snake species that lives underground feeds to a great extent on earthworms and soft-bodied insect larvae and slugs. A nonthreatening snake, the Eastern Wormsnake, when held in your hand, will use the spine on its tail to push itself forward, trying to bury its pointed head into the cracks between your fingers. Life underground also provides protection from surface predators such as birds, woodland snakes, and small mammals that would readily take an Eastern Wormsnake. Small mammals, such as shrews or voles, which frequent subsurface tunnels, are potential predators.

Eastern Wormsnakes are found in seemingly contrasting habitat types. They colonize habitats with loose, sandy soil, typically pine barrens ecosystems. They also are found in areas with exposed fractured bedrock or boulders that create openings for the snakes to retreat underground. And they are found in mesic hardwood forests with thick duff layers. The common factor is that they are often found in a habitat with a substrate that makes it easy to retreat underground and in the small, open-canopied or grassland areas associated with woodlands. Maybe "often" is not a good word to use with this species. Finding them in the Northeast is not particularly easy and almost always involves searching under many cover objects, rocks, logs, or refuse discarded by humans. Eastern Wormsnakes can be found on the surface but usually only at night, when the surface soil, vegetation, and leaf litter are not bone dry. On a warm, overcast day when a light rain or mist fills the air, they may also be moving on the surface. But it is more likely that they are just under the leaves, waiting for darkness. My most surprising encounter with an Eastern Wormsnake was finding a young one on the surface near a Timber Rattlesnake den on an oppressively hot, humid, misty, early afternoon in late September.

Northern Black Racer
Coluber constrictor constrictor

Type specimen described by Carolus Linnaeus in 1758, collected from the vicinity of Philadelphia, Pennsylvania

Total length: at birth = 10.1 inches (25.6 cm) to adult = 73.0 inches (185.4 cm)

Endangered in Maine

Threatened in New Hampshire and Vermont

As an adult, the Northern Black Racer is a long, slender, cylindrically shaped, smooth-scaled, shiny black snake. In contrast, young snakes of this species have a spotted dorsum, as do the similar-looking young of the Eastern Ratsnake, but the spots do not extend onto the tail in the Northern Black Racer as they do on the Eastern Ratsnake. The spotting pattern is lost by the time the snake is about 3 years old and about 30 inches (76 cm) long. As the name implies, the Black Racer is a fast-moving snake that spends most of its time in open grassy areas and in areas with open-canopied forests. They are most often encountered during the day, and when approached they quickly flee. They are feisty snakes, however, and do not hesitate to strike and bite repeatedly when threatened or cornered.

Although its scientific name is *Coluber constrictor constrictor*, the Northern Black Racer does not kill its prey by constricting. It will quickly coil around an escaping prey to pin it to the ground and them consume it live. It is an active hunter rather than an ambush hunter. While hunting, the Northern Black Racer moves through the grass with its head and neck elevated above the ground, apparently relying on eyesight for finding prey. Its preferred food items are extremely varied, comprising everything from moths, butterflies, and other invertebrates to salamanders, frogs, lizards, snakes, birds, and small mammals. Black Racers are able to take small Timber Rattlesnakes, a positive attribute to some people.

The Northern Black Racer, along with several other northeastern snakes, vibrates its tail rapidly when startled or threatened. If the tail comes in contact with dry grass or small branches, the sound is like that made by a rattlesnake or perhaps an insect. Is the snake warning potential predators to leave it alone? Or is it trying to attract a curious ground-feeding bird or small mammal to come close enough to capture? The jury is still out on that one, but what is known is that the snake does not hear the sound it makes, certainly not as mammals and birds do. Snakes have no external ear, but they are not completely deaf. They do have vestiges of an inner ear that is attached to their jaw bones. Through this they are sensitive to vibrations of the ground, such as an approaching predator or small prey, and perhaps they can also sense low-frequency airborne sounds. But snakes do not use sounds to communicate with other snakes. The tail rattling and the hissing sounds are for communicating to other species that do have an external ear.

Eastern Ratsnake
Pantherophis alleganiensis

Type specimen described by John Edwards Holbrook in 1836, collected from the summit of Blue Ridge, Virginia

Total length: hatchling = 10.2 inches (25.8 cm) to adult = 101.0 inches (256.5 cm)

Endangered in Massachusetts

Threatened in Vermont

Formerly known as the Black Rat Snake (*Elaphe obsoleta*), the Eastern Ratsnake is one of two large "black snakes" found in the Northeast, the other being the Northern Black Racer (*Coluber constrictor constrictor*). The genetic studies that placed the Eastern Ratsnake in the genus *Pantherophis* also divided what was considered the Black Rat Snake into three species, with one of the other species, *Pantherophis spiloides*, the Gray Ratsnake, presumably occupying portions of western New York, Pennsylvania, Maryland, Virginia, and West Virginia. More work is needed to map precisely where the transition from one species to the other actually occurs.

The patterning on the Eastern Ratsnake changes dramatically from hatchling to mature adult and has confused and bewildered many a casual observer. The young have a spotted form, with brown or grayish blotches that at first glance are reminiscent of the Eastern Milksnake. The Eastern Milksnake has brownish-red to red dorsal spots and smooth scales, whereas the Eastern Ratsnake has weakly keeled scales. The spots on the young Eastern Ratsnake extend onto the tail, but not onto the head. On the Eastern Milksnake, the spots appear on both the tail and the head. The spots on the Eastern Ratsnake slowly disappear by the time the snake is about 30 inches (76 cm) in total length. The Gray Ratsnake tends to keep juvenile patterning even as a full grown adult.

The Eastern Ratsnake is one of the most arboreal snakes. Their excellent tree-climbing ability is related to several factors. The first is that their bellies are flat, which gives them a better grip when climbing up a tree. In cross section, they look like a loaf of bread. The second factor is that they climb better when the tree has rough bark rather than smooth bark. And the third is based on their internal anatomy and physiology. The length-to-diameter ratio, for snakes that are good climbers, points to the best climbers being those that are proportionally more slender. But switching from a basically horizontal lifestyle to one where they must cope with the effects of gravity means that things must change internally as well. Gravity affects the snake's ability to circulate blood when its body is vertical. The snake that climbs will have its heart closer to the head, whereas the nonclimbing snake will have its heart closer to its midpoint. An Eastern Ratsnake can climb considerable distances without the aid of branches and therefore has its heart closer to its head than most snakes. Snakes that climb just a little will need side branches to aid their climb, a place to return to a more horizontal position.

Eastern Ratsnakes are predators that secure their prey by constricting. As the name implies, they eat a lot of rodents and can take prey as big as an Eastern Gray Squirrel. With their climbing ability, they can also ascend trees and take birds or bird eggs from nests high above the ground. They may also climb the posts that support bird houses or bird feeders to capture prey.

Eastern Milksnake
Lampropeltis triangulum triangulum

Type specimen described by Bernard-Germain-Etienne de le Ville-sur-Illon Lacèpede in 1788, collected from the vicinity of New York City

Total length: hatchling = 6.7 inches (17.0 cm) to adult = 52.0 inches (132.1 cm)

The Eastern Milksnake is a beautiful snake. It is marked with a row of large, roundish, red or reddish-brown spots with black borders on its back alternating with smaller ones on its sides. The elongated spot on the head and neck surround a gray to tan V- or Y-shaped light-colored marking that matches the base color of the snake. Even the belly of the Eastern Milksnake is attractive, with a black-and-white checkerboard pattern. The Eastern Milksnake is found throughout the region except for northern New England.

It is unfortunate that the Eastern Milksnake is a spotted snake, because so many people believe that a spotted snake is a venomous snake. In many areas it is referred to as the "spotted adder," an unfortunate misnomer. The spots run all the way to the tip of the tail, which leads uninformed observers to think the alternating red spots with white spaces between them is the "rattle" of a rattlesnake, which in truth it resembles not in the least. The tail of an Eastern Milksnake is long and slender, whereas the tail of a rattlesnake ends in a broad, flattened segmented rattle. When the Eastern Milksnake is threatened, it may raise its tail and vibrate it vigorously, sometimes contacting dry leaves, a branch, or other solid object, resulting in the observer hearing a rattling sound. For people who fear snakes, especially venomous snakes, the Eastern Milksnake's fate is sealed. It must die, a great loss of one of our most beneficial snakes. This species does not present a real threat to people or their animals. It does present a real threat to the small rodents—mice, voles, and rats—found near people's dwellings. It is these small rodents that can damage structures built by humans and carry diseases that present a danger to humans, not the snakes that help keep the small mammal populations under control.

Eastern Milksnakes are constrictors. They kill their prey by grasping them with a quick strike and then holding them with their teeth while wrapping their bodies around the victim. The snakes sense the prey's heartbeat to determine when it is dead. The prey is then swallowed whole after it is dead. It was commonly believed until recently that constrictors killed their prey by crushing or suffocating them. That is, after all, what it looks like. It is now known that, at least for some constrictors, there is more to the story. The wrapped coils of the snake shut off blood flow and oxygen to the brain, heart, and other vital organs of the prey. The snake essentially applies a tourniquet to the body of its prey, which quickly leads to the prey losing consciousness, followed by death in a matter of seconds.

Smooth Greensnake
Opheodrys vernalis

Type specimen described by Richard Harlan in 1826, collected from Pennsylvania or New Jersey

Total length: hatchling = 3.3 inches (8.4 cm) to adult = 26.0 inches (66.0 cm)

The small, slender, smooth-scaled Smooth Green-snake is stunningly beautiful, with a vivid, almost grass-green dorsum and pale yellow belly. Sometimes called the grass snake, when it is lying motionless in a grassy field or lawn, you might miss it unless it moves, and then it moves swiftly. A sharp eye is needed to keep track of its movements as it disappears among the bases of the grassy vegetation. It is not a great climber, but it can be found basking or searching for food in low branches and shrubs. The Smooth Greensnake can also be found hiding under rocks along road cuts, where its green color contrasts with the dull gray and brown rocks. It does not immediately flee when exposed by rock lifting, allowing time for a photograph or two before carefully returning the rock to its previous position. In addition to road cuts, I generally look for it in small woodland openings with sparse grass cover or small glades where terrestrial sedges are the dominant ground cover.

The young, slightly larger than the diameter of a pencil lead, are not as brilliant a green as the adults, tending more toward the dark olive drab or bluish-green shades. It takes a while to develop the vivid green of an adult. The adult Smooth Greensnake is not "green" at all, except to our eye. Its remarkable color is a mix of yellow pigments overlaying blue pigments that become obvious when the snake dies. The yellow pigment disintegrates first, leaving a blue, smooth-scaled specimen that will make you wonder whether the dead snake you found might be a new species.

The Smooth Greensnake is oviparous; that is, like many snakes, it lays eggs rather than giving birth to live young. It lays small clutches of four to six eggs under rocks, logs, and boards that hatch in late August or early September. What is unusual about their nesting is that the length of incubation is the shortest of any snakes and that the time after being laid varies tremendously. The eggs could hatch in less than 4 days to as long as 23 days. The female Smooth Greensnake can regulate the amount of time the eggs are in her body. During that period the embryos continue to develop. Because increased temperature speeds up the development of the embryo, she basks such that her own body temperature rises and aids in transfer of heat to the eggs. By the time she lays the eggs, the embryos are far advanced.

The Smooth Greensnake is a bit of a picky eater, preferring hairless caterpillars to most other food, although it will take spiders and other small soft-bodied insects. While not likely to bite a handler, this characteristic makes the Smooth Greensnake a poor captive, rarely taking food and exhibiting a nervous disposition when in a terrarium.

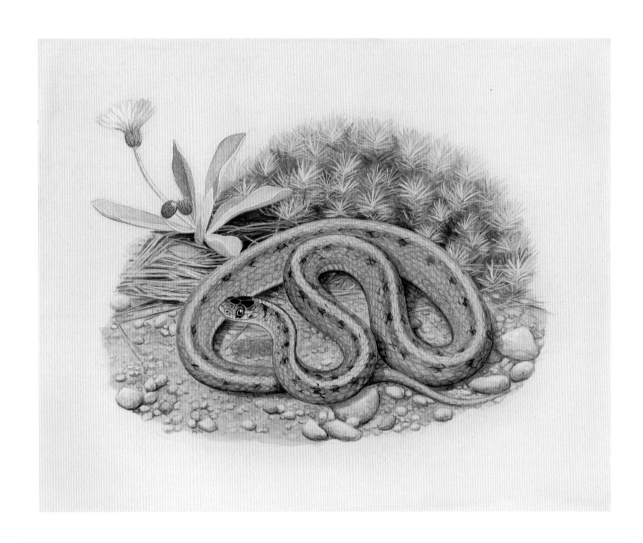

Northern Brownsnake
Storeria dekayi dekayi

Type specimen described by John Edwards Holbrook in 1836, collected near Cambridge, Massachusetts

Total length: at birth = 2.8 inches (7.0 cm) to adult = 19.4 inches (49.3 cm)

The Northern Brownsnake, known to many older naturalist as DeKay's Snake, was named for two early nineteenth-century naturalists, David Humphreys Storer of Massachusetts and James Ellsworth DeKay of New York, making it the only snake in North America whose genus and species names honor people. Both Storer and DeKay were trained professionally as medical doctors, a somewhat common situation for many early naturalists, but for both, natural history was their passion. DeKay has been credited with collecting the first specimen of this species, which he found on Long Island near where he lived and was later named for him by his friend John Edwards Holbrook, who was also a physician-naturalist.

As an adult, this small brown snake with keeled scales is most frequently mistaken for a young Common Garter-snake or an adult Eastern Wormsnake. The first characteristic to look for is the scales: wormsnakes have smooth scales. Next, check for a uniformly colored light stripe down the back bordered by two parallel rows of black dots and a checkerboard pattern on the sides, which is most obvious if the snake is gravid or has just fed. Otherwise the pattern can be seen by gently stretching the skin. If still in doubt, look for the dark vertical bar on each side of the head to confirm the Northern Brownsnake. Female Northern Brownsnakes are ovoviviparous, that is, they give birth to live young. These wire-thin neonates can be confused with newborn Ring-necked Snakes, as they both have light-colored collars around their necks. If the scales are keeled, it is a Northern Brownsnake, if smooth, it is a Ringed-necked Snake.

Northern Brownsnakes, locally called city snakes or garden snakes, are tolerant of development and survive quite well in city parks, rural gardens, and cemeteries—almost any place where there is suitable cover and a food supply. They are known to occur within the city limits of Wilmington, Delaware, and Central Park in New York City. Rocks, logs, and piles of debris are suitable cover.

Northern Brownsnakes are active hunters rather than ambush predators like most snakes that prey on small mammals and birds. The food preference of Northern Brownsnakes consists of snails, slugs, and earthworms, but they will take insects, sow bugs, and other crunchier kinds of invertebrates. Like all snakes, they swallow their food whole and obviously can't take large snails with their hard shells. But Northern Brownsnakes have evolved a solution to that dilemma. After grabbing a snail by the soft portion of its body, a Northern Brownsnake can extend its lower jaw, which is not fused in the front into one piece, to get a good grip and pull the snail out of its shell. Only the Northern Brownsnake and its cousin the Northern Red-bellied Snake have this ability.

Short-headed Gartersnake
Thamnophis brachystoma

Type specimen described by Edward Drinker Cope in 1892, collected along the
Allegheny River near Franklin, Venango County, Pennsylvania

Total length: at birth = 4.9 inches (12.5 cm) to adult = 22.0 inches (55.9 cm)

The Short-headed Gartersnake is the shortest of northeastern gartersnakes and has the smallest range. They are found primarily in the Allegheny Plateau portion of western Pennsylvania and western New York, in an area that is almost perfectly defined as the Allegheny River watershed. One outlier in south-central New York falls within the Susquehanna River watershed, and debate has continued for years as to whether this population is natural or introduced. Given appropriate habitat, this species will readily become established in areas where it is released. Populations have been introduced and established in Pennsylvania in Erie County, northwest of its natural range, south of its natural range in Butler County and the Pittsburgh area, and possibly in Mercer County at the western edge of its range.

The Short-headed Gartersnake has three distinct yellow to tan stripes along the length of its body. An attractive snake, its back between the three lateral stripes is a dark chocolate brown, and its sides below the lateral stripes more of a milk chocolate brown. The checkerboard pattern often seen on the Common Gartersnake is not found on the Short-headed Gartersnake. The short, narrow head is the same width as the neck, a body shape reminiscent of fossorial snake species.

This is a species of the open fields and meadows bordered by woodlands but the Short-headed Gartersnake does not wander far under a full canopy. Its diet is predominantly earthworms, which it actively hunts. When startled or handled, it will thrash about vigorously at first and spread excrement on the predator—or a person who thinks it's okay to pick it up—but it never attempts to bite as would other gartersnakes.

Short-headed Gartersnakes are most often found under some sort of cover object, such as flat stones, discarded boards, or around fallen barns or outbuildings. The best place I have found to search for this small serpent is in shale road-cuts bordered by open grassy slopes. Cemeteries where the grass has been allowed to grow tall and abandoned agricultural fields provide suitable habitat, but as the open-growth saplings are replaced through natural succession to closed-canopy forests, the population of Short-headed Gartersnakes declines. The general consensus from those who have monitored its populations for decades is that it still occurs at virtually all the places where it used to occur, but its numbers have dropped considerably since the 1950s as woody vegetation has replaced grassland areas.

Common Gartersnake
Thamnophis sirtalis

Type specimen described by Carolus Linnaeus in 1758, collected from Canada
Total length: at birth = 4.9 inches (12.5 cm) to adult = 48.7 inches (123.8 cm)
State reptile of Massachusetts and Virginia

The Common Gartersnake is found virtually everywhere in our region. Twelve subspecies have been described, two of which occur in the Northeast: the Eastern Gartersnake (*Thamnophis sirtalis sirtalis*) and the Maritime Gartersnake (*T. sirtalis pallidus*). The Common Gartersnake has the most extensive range of any snake in North America, being found from coast to coast, south to the southern limits of the United States, and farther north, deep into Canada, than any other snake species in the Western Hemisphere. A diurnal species at home in open grasslands, woodlands, cultivated fields, swamps, dry habitats, lawns, gardens, and city parks, this is the snake seen most frequently by people.

With such a large range and large number of subspecies, the gartersnake's color pattern is quite variable, as one might expect. One consistent characteristic is that all gartersnakes have keeled scales. In the Northeast, most individuals have three yellow stripes on a dark black or brown to olive body, although some individuals have reduced stripes or may even lack stripes completely. Between the stripes is often a checkerboard pattern of light and darker square-shaped spots. Full-grown females are significantly larger than the males. The maritime subspecies, which is found in northern New England as far south as Boston, is smaller than the eastern subspecies, with a maximum length of 36.1 inches (91.7 cm).

Common Gartersnakes often den communally during the winter. They are the first snake to emerge in the spring and the last to enter hibernation in the fall. Shortly after emerging in the spring, they may be observed congregating in "mating balls," with one female being accompanied by many smaller males that used a scent trail left by the female to locate her. A female can produce a litter of 30 young or more. Researchers note that smaller females tend to produce litters of young that are biased in favor of females, whereas larger females produce more young males.

This cosmopolitan species also has a cosmopolitan diet. It preys on numerous invertebrate species as well as both adult and larval amphibians, fish, small mammals, and even birds. In many areas, worms or other soft-bodied invertebrates and amphibians make up the bulk of the Common Gartersnake's diet. Often referred to as the Garden Snake, it is a good species to have around if you are a gardener—or simply a homeowner—to help keep pest species under control.

Smooth Earthsnake
Virginia valeriae

Type specimen described by Spencer Fullerton Baird and Charles Frederic Girard in 1853, collected from Kent County, Maryland

Total length: at birth = 3.1 inches (8.0 cm) to adult = 15.5 inches (39.3 cm)

Endangered in Maryland (only applies to subspecies *V. valeriae pulchra*, the Mountain Earthsnake)

The Smooth Earthsnake is a small, nondescript gray to brown, stout-bodied, fossorial species. A close examination might reveal a series of tiny black dots on the dorsum. It has both smooth and slightly keeled scales and a small, pointed head with no discernible neck. One of the smallest snakes in the Northeast, females reach sexual maturity at only 7.4 inches (18.5 cm) and males at only 4.9 inches (12.5 cm). There are two species with which it might be confused: the Northern Brownsnake, which has strongly keeled scales, and the Eastern Wormsnake, which has completely smooth scales. The specific name *valeria* is to honor Miss Valeria Blaney, who was the first to collect this species.

The Smooth Earthsnake occurs in the Northeast as two allopatric subspecies, that is, as occurring in separate nonoverlapping geographical areas. The Mountain Earthsnake (*V. valeriae pulchra*) is endemic to the Northeast, being restricted to a small band extending from just south of the New York border in Pennsylvania through the western panhandle of Maryland into West Virginia and extending just across the border to one location in Highland County, Virginia. The Eastern Smooth Earthsnake (*V. valeriae valeriae*) is more widespread, especially in the southeastern states. In our region, it is found from central New Jersey south, primarily east of the folded Appalachians. It is also found in western West Virginia and Virginia. The Eastern Smooth Earthsnake has most likely been extirpated from eastern Pennsylvania by extensive development for housing, industry, and agriculture, but the Mountain Earthsnake still survives in the western part of the state.

The Smooth Earthsnake inhabits a wide variety of habitats, including grasslands; pastures; both urban and suburban woodlots; and hardwood, pine, and mixed forests. At some sites it is locally common. Shortly after a rainstorm is the best time to search for Smooth Earthsnakes in leaf litter, under rocks and logs, or under any type of trash or debris discarded by humans. Earthsnakes will take slugs, soft-bodied insects, and their larvae but prefer to feed almost exclusively on earthworms. One would expect that with such broad habitat requirements, and a food source readily available almost everywhere, the species would be better studied, but many of the locations where it has been observed or collected are widely disjunct both in time and space. Its secretive nature leaves much to learn, including a clearer picture of its distribution that could well be more extensive than is currently known.

4 Wicked Big Puddles

SEASONAL WETLANDS have far more significance to wildlife than their size might indicate. Nearly 40% of northeastern herps—more than 60 species—use seasonally flooded wetlands during some part of their annual cycle. Most of these wetlands are quite small, a few acres constituting a large one. But the biodiversity they support coupled with a seasonal abundance of amphibian eggs and larvae make them a main contributor to energy flow in the surrounding natural communities.

Naturally occurring seasonal wetlands form in slight depressions. They have no permanent outlet, often no inlet, and are generally termed "isolated wetlands." Their source of water comes from rain, snowmelt, spring runoff, floods, or a perched water table. As the name implies, they are part-time wetlands. The water comes and goes on an annual basis, so the size of the pool and depth of the water also vary greatly throughout the year. Those that fill in the spring from snowmelt and rain, in time for amphibian migration to breeding ponds, are properly called vernal pools. Around the vernal equinox, March 21 more or less, when daylight hours are nearly equal to nighttime hours and the frost is leaving the ground, is the time of the year when many of the frogs, toads, and salamanders are once again becoming active with a strong urge to breed. There are also many seasonal wetlands that fill in the fall rather than the spring, yet few people call them autumnal pools. A landscape populated with seasonal pools nestled within the forest provides ideal habitat for these species as they make their annual migration to the breeding ponds.

Seasonal wetlands go by lots of other names—ephemeral wetlands, pocosins, Carolina bays, low swales, and seasonal floodplain pools, to name a few—but my favorite is "wicked big puddles," a name given to these pools by Leo Kenney of the Vernal Pool Association and used extensively in a public school curriculum in Massachusetts since the early 1980s.

There are a number of unnatural seasonal wetlands that various seasonal wetland–dependent species utilize. Ditches along roads and ruts made by off-road vehicles such as logging skidders form linear "wicked little puddles" that may hold water long enough for some species to successfully breed. But that is a perilous existence. Roadside ditches by their very nature contain road salts and numerous chemicals spewed by motorized vehicles. And species that try to breed in woodland ruts may have their lives ended by additional off-road traffic.

A key feature of seasonal wetlands is the length of hydroperiod, that is, how long they hold water. It is of little value to wetland-breeding amphibians if water is not present from the time the adults breed and lay eggs until the larvae mature and transform into the terrestrial form. Seasonal wetlands become a sink for species that can't complete all these stages before the wetland dries up; the conditions lure them to breed in less-than-suitable wet areas, giving the larvae little or no chance of maturing. The effort by the adults to reproduce is lost.

Complete drying of these pools provides another benefit. A wetland that does not completely dry up accumulates dead leaves and other organic material that decompose in an oxygen-depleted environment. If the pool dries completely, oxygen in the atmosphere speeds the decomposition process on the former pool bottom. When the pool refills, there will be a higher concentration of dissolved oxygen in the water column than in an isolated pool that rarely or never dries completely. Dissolved oxygen in the water column is important to amphibian larvae that "breathe" through their gills. The dissolved oxygen levels in isolated pools is still far less than in a flowing stream, so salamander larvae in pools have external gills that are significantly bushier than salamander larvae found in headwater streams.

Each year, usually by late summer, the pool goes dry, but some seasonal pools may not dry out completely every year. A pool that has a hydroperiod of less than a year will not support amphibians that have a larval period of two years or more. Bullfrogs, Green Frogs, and Two-lined Salamanders are some of the species that cannot successfully breed in seasonal wetlands. The good news is that fish—predatory fish—cannot survive in an isolated pool that dries up on a yearly basis. Although there certainly are other aquatic predators of amphibians, their eggs, and their larvae, eliminating fish as one of those predators significantly increases survival rate. Predatory aquatic insects, wandering Spotted Turtles, and Red-spotted Newts do prey on amphibian eggs and larvae in seasonal pools.

Species that breed only in seasonal wetlands are classified as either obligate species or facultative species. A species that can only breed in a seasonal wetland is an obligate species; one that can breed in these wetlands but can also successfully breed in other kinds of wetlands is known as a facultative species. Ambystomid salamanders (salamanders in the genus *Ambystoma*), Wood Frogs, and Spadefoot Toads are considered obligate species, whereas Spring Peepers, Gray Treefrogs, American and Fowler's Toads, Four-toed Salamanders, Red-spotted Newts, Spotted Turtles, and a host of aquatic insects are facultative species. Although ambystomid salamanders and Wood Frogs breed most frequently in seasonal wetlands, I have found Spotted Salamanders, Eastern Tiger Salamanders, Jefferson Salamanders, and Wood Frogs breeding in permanent wetlands such as beaver ponds, floodplain wetlands, and sometimes even in riverine systems with fish present.

A truly seasonal wetland obligate species is the Fairy Shrimp (*Eubranchipus* spp.), a small, 0.5- to 1.5-inch (1.3- to 3.8-cm), soft-bodied crustacean. As the pool dries out, the female produces "winter eggs" that remain in the organic material on the bottom of the pool and dry out as the pool dries out. The eggs hatch when the pool refills, providing prey for both the adult salamanders and their maturing larvae.

For seasonal pool–breeding salamanders, forest surrounds the best pools. Trees that are typically closest to the wetland, or sometimes within it, are elm, ash, and Red Maple. The canopy helps moderate the pool temperature, which in turn helps maintain adequate dissolved oxygen levels. Trees that fall into the pool basin provide sturdy branches on which the salamanders can attach their egg masses. For pools in more open areas, the pool often becomes colonized by herbaceous vegetation, which provides hiding places for the larvae. Woodlands should be nearby, however, because the adults of these seasonal pool–breeding amphibians spend most of their life in the forest foraging and overwintering there.

Blue-spotted Salamander
Ambystoma laterale

Type specimen described by Edward Hallowell in 1856, collected
from Marquette, Michigan

Total length: metamorph = 2.0 inches (5.2 cm) to adult = 5.5 inches (14.0 cm)

Endangered in New Jersey and Pennsylvania

Threatened in Connecticut

The Blue-spotted Salamander is the smallest ambystomid salamander. It is approximately the same length as the stockier Marbled Salamander, but a more slender body means the Blue-spotted Salamander has less mass. Positive identification is difficult because this species hybridizes with the Jefferson Salamander, a phenomenon that is discussed in more detail in the Jefferson Salamander account below.

A pure diploid Blue-spotted Salamander is marked with baby blue or sometimes whitish or silvery flecking on its back and sides. Its ventral surface is uniformly dark, nearly as black or blue black as the ground color of its dorsum. When threatened, the Blue-spotted Salamander will raise its tail and wave it in an undulating manner while exuding a white, sticky, foul-tasting slime in an effort to repel a potential predator.

The forest is the adult Blue-spotted Salamander's home for 11 months of the year. Most of its time is spent underground or hiding under logs and rocks, coming to the surface only when wet or damp conditions prevail. But the central feature of the Blue-spotted Salamander's home range is a fish-free seasonal pool. It breeds in these pools shortly after snowmelt and sometimes while ice is still clinging to the edges and tussocks in the wetland. To reach the breeding pool, the Blue-spotted Salamander must complete its nocturnal migration from its overwintering site in the surrounding upland forest. This journey, which may be more than a half mile, becomes hazardous if roads dissect the migratory route. Moving at night, with a dark color that blends into the asphalt of the roadway, the Blue-spotted Salamander is not nearly as obvious as its larger cousin, the Spotted Salamander. Unnoticed, many Blue-spotted Salamanders end up as roadkill, even in areas where drivers swerve to avoid the larger, Spotted Salamander with its bright yellow spots.

In the pond, the Blue-spotted Salamander's egg masses are attached in small, loose clumps of two to thirteen eggs to twigs or other stiff vegetation below the water surface. It takes three to four weeks for the eggs to hatch and gilled larvae to emerge. During that period, if the water level drops so that the eggs are above the water surface, survival is greatly decreased and may drop to zero. The larvae that survive this first threat still face the threat of predatory aquatic insects. If all goes well, the larvae mature in two to four months, lose their gills, and move out into the surrounding forests. This late-summer exodus from the breeding pools may result in a second peak of roadkills when the recent metamorphs attempt to cross the road. For a metamorph to become a sexually mature adult may take another two growing seasons.

Jefferson Salamander
Ambystoma jeffersonianum

Type specimen described by Jacob Green in 1827, collected from Cannonsburg, Washington County, Pennsylvania

Total length: metamorph = 1.9 inches (4.8 cm) to adult = 8.3 inches (21.0 cm)

The Jefferson Salamander was named for the college of the same name in honor of the third US president. In 1865, this college joined together with another nearby college named for the first US president, forming Washington and Jefferson College, southwest of Pittsburgh. I am sure President Jefferson, the naturalist, would have been pleased to have his name attached to a salamander, albeit an indirect connection.

The Jefferson Salamander is a medium-sized, comparatively slender ambystomid salamander with long toes. Its ground color is light to chocolate brown to brownish gray with a lighter pale belly. Its sides are marked with blue, silvery, or whitish flecking, the flecking being smaller and less distinctly blue than normally seen on a pure Blue-spotted Salamander.

Prior to 1856, specimens of Jefferson and Blue-spotted Salamanders were not recognized as separate species. Even into the mid-1900s, the literature did not always separate the two species. To confuse things even more, two additional species were named that appeared to be intermediates between Jefferson and Blue-spotted Salamanders: the Silvery Salamander (in 1867) and Tremblay's Salamander (in 1943). Subsequent investigations demonstrated that Silvery and Tremblay's Salamanders were triploids rather than diploids; that is, they had three sets of chromosomes instead of the normal two sets.

Those two latter forms are no longer considered to be discrete species but rather are part of a species swarm, referred to as the *Ambystoma jeffersonianum* Complex or the Jefferson–Blue-spotted Salamander Complex. This complex includes the Jefferson Salamander, the Blue-spotted Salamander, the two triploids, and several other combinations of the two parent species, which includes both diploid and tetraploid hybrids.

The diploid Blue-spotted Salamander has two sets of chromosomes that are referred to as LL. Similarly, the diploid Jefferson Salamander is referred to as JJ. The triploid Silvery Salamander, which looks just a bit more like a Jefferson Salamander than a Blue-spotted Salamander, has three sets of chromosomes (LJJ), and likewise the triploid Tremblay's Salamander, which is more similar in appearance to the Blue-spotted Salamander, has an LLJ genotype. Both of these triploid forms are nearly all females, with less than 2% males. The few males that are produced are sterile. Diploid LJ salamanders plus tetraploid LJJJ, LLJJ, and LLLJ salamanders have also been identified as part of this complex.

In most instances where cross-species hybrids are produced, the offspring are sterile, as seen in mules, which are the result of breeding a male donkey and a female horse. Not so with these salamanders. The diploid, triploid, and tetraploid hybrids can all produce viable offspring. The female hybrids do pick up a spermatophore, or sperm packet, enabling the sperm to contact the surface membrane of the egg without penetrating it. This action is enough to stimulate the egg to begin to develop into a viable embryo. Without the incorporation of the sperm to fertilize the egg, the offspring from this reproductive encounter lead to a new generation of all-female salamanders.

This reproductive strategy challenges the traditional definition of a species, as well as the formation and viability of hybrids, and offers a lesson in evolutionary mechanisms. It also challenges conservation biologists, who must decide whether efforts to protect certain populations are an effective way to allocate limited resources, such as the disjunct populations of pure Blue-spotted Salamanders on Long Island, New York, which are completely isolated from any genetic material from Jefferson Salamanders.

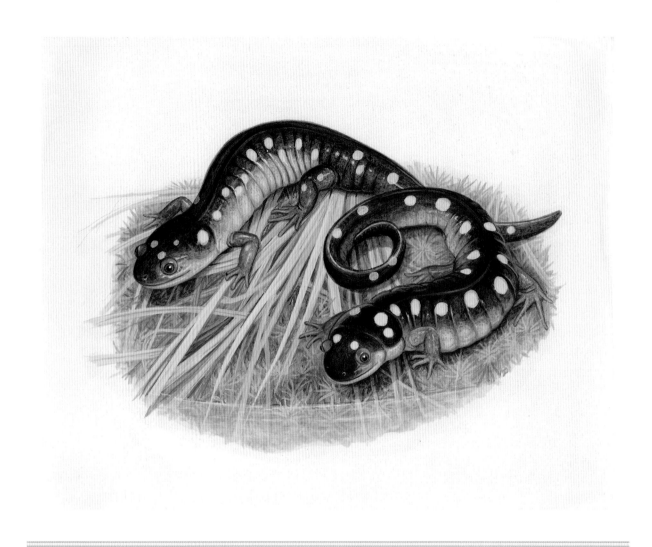

Spotted Salamander
Ambystoma maculata

Type specimen described by George Shaw in 1802, collected from the vicinity of Charleston, South Carolina

Total length: metamorph = 1.1 inches (2.7 cm) to adult = 9.8 inches (24.8 cm)

This large, stocky salamander should not be confused with any other salamander in its range. The Spotted Salamander's black back and grayish belly are marked with two irregular rows of circular, bright yellow, or somewhat orange spots running the length of its body and tail. A Spotted Salamander can live for more than 20 years.

Male Spotted Salamanders begin their spring breeding migration several days before the females. During breeding season, the sexes can be differentiated if examined closely. Males are slimmer than females, who by this time of year are carrying hundreds of well-developed eggs. Males have a swollen vent or cloaca, the opening on their ventral surface at the base of the tail, through which they excrete digestive waste and also release spermatophores from their reproductive tract. The female's cloaca is much less swollen if it is swollen at all. Warm rains or at least wet surfaces trigger the movement to the breeding pools. If the weather does not cooperate—that is, extended cold or dry periods occur after the males make their migration—the males may end up waiting in the pools for days or possibly weeks for the females to arrive.

When the females do arrive, the excitement begins. Males congregate around receptive females, sometimes in large numbers, and perform a courtship dance. The male deposits spermatophores on the pool bottom or on twigs. The spermatophore appears as a tiny little pedestal with a whitish tip where the sperm is held. While the breeding activity occurs mostly at night, these sperm packets can be seen during daytime hours, indicating that the breeding season has started even if the adults are not seen. The female positions herself over the spermatophore and takes up the sperm into her cloaca. Thus fertilization of the eggs occurs internally. Over the next day or two, the females deposit eggs in a globular mass of 100 or 200 eggs attached to twigs or other firm vegetation below the water surface. Initially, the eggs are pinhead-sized black dots. The eggs absorb water over the next several days until they are a mass of pea-sized eggs surrounded by a firm, grape-sized gelatinous mass. The Spotted Salamander egg mass at this stage can be distinguished from all other ambystomid egg masses by its much firmer feel. Females usually deposit more than one egg mass, and toward the end they produce egg masses with fewer eggs. Water temperature affects how long it takes for embryos to develop, but larvae can be expected to hatch in 30 days or so.

The Spotted Salamander is the iconic mass-migrating, vernal-pool-breeding amphibian. It is found throughout the Northeast and has become the poster child in the region for amphibians that are being killed on the road each spring as they make their breeding migration. In some areas, considerable numbers are killed on roads by vehicles each spring as they move from upland overwintering sites to wetland breeding habitats. It is not unusual to see bands of citizens walking the roadways at night with flashlights counting the dead and helping the living make it across the roads. A growing number of amphibian tunnels are being installed under roadways so that migrating salamanders and frogs do not have to face the hazard of trying to cross the road.

Marbled Salamander
Ambystoma opacum

Type specimen described by Johann Gravenhorst in 1807, collected from New York
Total length: metamorph = 1.7 inches (4.4 cm) to adult = 5.0 inches (12.7 cm)
Endangered in New Hampshire
Threatened in Massachusetts

The Marbled Salamander is a black salamander with white hourglass-shaped crossbands that make its back look like it could be a white salamander with a line of large black vertebral spots. The dorsal markings are white or silvery white on the male but more of a dull gray on the female. The belly and legs are black, as is the rest of the animal. It is a short, stocky species with a short, stocky tail and proportionately shorter legs than most of the other ambystomids. The white of the male is at its crispest during breeding season.

The Marbled Salamander is the only late-summer- or fall-breeding ambystomid salamander. It may select the same pools in which other ambystomid salamanders breed. But the female doesn't actually lay her eggs in the water. Instead, she shows up at the breeding pond when the water level is low or even dry. She deposits her eggs under a rock, log, or other debris on land, not far from the water's edge, but in a location where rising waters that follow a late-season rain will submerge the eggs. The female remains with and guards the eggs until they are flooded. The embryos begin developing before being submerged. The rate at which the embryos develop is inversely correlated to the amount of moisture present. Under ideal conditions, the eggs may be ready to hatch in as little as 15 days. Under drier conditions, it may take nearly two months for the embryos to mature within the eggs. When the rains finally come, the eggs may hatch within a couple hours or take several weeks, depending on their stage of development. If the rains do not come in time, the entire reproductive effort for that pond is lost that year.

In ponds that fill with rainwater, the larvae survive, growing little over the winter. But they will be larger than recently hatched Spotted, Tiger, or Blue-spotted Salamander larvae that hatch in late winter or early spring. Marbled Salamander larvae will be able to prey on these other larvae. But the tide will turn; the other ambystomid larvae will grow faster, and the prey species will become the predators. By the time the Marbled Salamander larvae are mature and ready to transform into adults, eight or nine months may have passed, making their larval development the longest of any ambystomids, as the other species develop into metamorphs within just one growing season, a period of perhaps five or six months. When the metamorphs emerge, they lack the distinct patterning of the adults and may be confused with Slimy Salamanders. The Marbled Salamander has 11 costal grooves, whereas the Slimy Salamander has 15 to 17 costal grooves.

American Toad

Anaxyrus americanus

Type specimen described by John Edwards Holbrook in 1836, collected from the vicinity of Philadelphia, Pennsylvania

Head–body length: metamorph = 0.3 inches (0.7 cm) to adult = 4.4 inches (11.2 cm)

The American Toad, formerly in the genus *Bufo*, has a huge range that extends from southern Labrador and Hudson Bay south to Louisiana and west to the Dakotas and Manitoba. It is a rough-skinned species marked with black spots on its chest, belly, and back. Within the larger black spots on its back are one or two orange or reddish-brown wart-like bumps. These characters are usually enough to identify the species, but where its range overlaps that of the Fowler's Toad, the differences are not always obvious, and sometimes the two may hybridize. The American Toad has spiny warts on the dorsal surface of its hind legs and the cranial crest, which separates the eye from the peanut-shaped paratoid glands. The cranial crest does not touch the paratoid gland, or if it does, only by the spur on the leg of the L-shaped crest.

If there is difficulty in visually confirming an American Toad, wait for its breeding choruses. It has the most musical chorus of any of the northeastern anurans, a name given to the group of amphibians that includes all frogs and toads. In spring, an individual male will launch into a 5- to 30-second trill and will soon be joined by other males in the breeding pond until the overlapping songs blend into what may become hours of musical, high-pitched trilling to attract the females.

When a receptive female is sensed, she is often surrounded by males all vying for a chance to mate. The female in the crowd is easy to identify, as females are significantly bigger than are the males. As in most anuran species, the successful suitor grasps the female in amplexus, a behavior where he uses his forelegs to hold her just behind her forelegs to stimulate the release of her eggs. This places him in position to deposit sperm to fertilize her eggs as she lays them in two long, parallel strings consisting of thousands of eggs. The strands of eggs remain obvious on the surface of the pond for several days, mapping the route the amplexing pair took during the breeding frenzy and often crisscrossing the route of other amorous pairs of toads.

After the eggs hatch within a week or so, the resulting tadpoles remain together in a swarm of tiny black dots with tails for a few days until they grow a bit and mature. At this stage they are vegetarians, using their five rows of tiny rasping teeth, two above and three below the mouth opening, to scrape the surface cells off any vegetation they encounter. It is only as the metamorphosing terrestrial form that they assume the life of a predator, consuming all types of invertebrates and small vertebrates that they can fit into their mouths. Emerging toad tadpoles are the smallest form of terrestrial frog in the Northeast. During its potential 30-year lifespan, an American Toad's body mass may increase more than 3,800 times from the smallest metamorph to the largest adult female. This size increase reflects the change in its predator role as the size of its prey changes dramatically during the toad's lifetime.

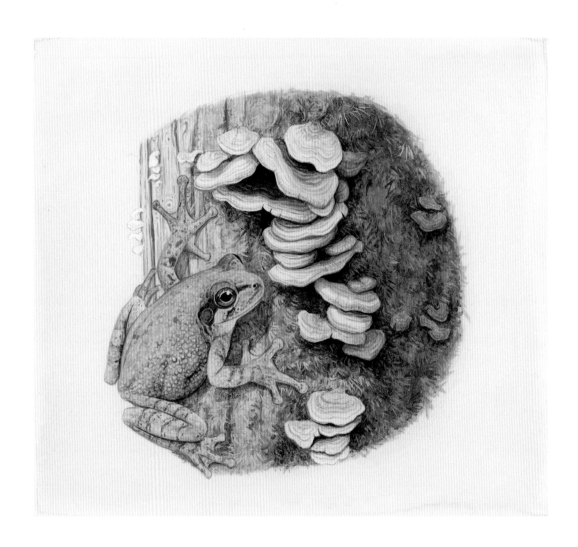

Gray Treefrog
Hyla versicolor

Type specimen described by John Eatton LeConte in 1825, collected from the vicinity of New York City

Head–body length: metamorph = 1.3 inches (3.2 cm) to adult = 2.5 inches (6.4 cm)

Endangered in New Jersey (Cope's Gray Treefrog)

The Gray Treefrog and its sibling species, Cope's Gray Treefrog (*Hyla chrysoscelis*), look identical on the outside. In fact, Cope's Gray Treefrog was not recognized as a separate species until 1880 after it was collected in Bathrop County, Texas. In the field, the two species can't be told apart in the broad area where there ranges overlap unless you hear them calling. In New York and New England, it is safe to call any treefrog observed or heard a Gray Treefrog; Cope's Gray Treefrog has not been found that far north. From Pennsylvania southward, it could be either species, so listen for the call.

Their call is a short, pleasant-sounding trill lasting less than a second and repeated 10 or 12 times per minute. The Gray Treefrog sings at about half the rate of Cope's Gray Treefrog, with just 16 to 35 notes per second compared to 29 to 64 notes per second. Temperature affects the rate of the trill in both species, with a slower trill on cool nights and a more rapid trill on warmer nights. The call might remind one of the trill of an American Toad, but the toad does not call in short bursts like the treefrog does. Rather, the toad trills for more than five seconds to half a minute at a time.

There is another difference separating the two species of Gray Treefrog. *H. chrysoscelis* is a diploid species with 24 chromosomes, whereas *H. versicolor* is a tetraploid with 48 chromosomes. A diploid is the normal condition for most plants and animals, that is, the individual has two sets of chromosomes. In a tetraploid, each individual has four sets of chromosomes. The greater number of chromosomes means the cell and its nucleus must also be larger. While certainly not a field character, measuring the diameter of the cell or the nucleus will enable you to determine which species you have, with *H. chrysoscelis* being about three-quarters the size of *H. versicolor*.

No matter which species you have, both species are almost completely nocturnal, moving during the day only in deep woods when the ground is damp from rain or mist. Treefrogs are excellent climbers, equipped with huge toe pads at the end of each digit that give them the grip to climb high in trees or cling to a glass windowpane. On rainy nights they cling to windows, feeding on the insects attracted to the lights from the home.

Both species are marked with a cryptic lichen pattern of various shades of gray with darker streaks and often a touch of green that blends in with the tree bark, making them difficult to find when calling from their elevated perch perhaps 7 to 20 feet (2 to 6 m) above the wetland. To make them even more difficult to observe, they have the ability to change their color, from gray to green, depending on the environmental conditions and the perch to which they are clinging. When their legs are extended, there is a bright yellow wash on the groin extending to the inner thighs. In contrast, recent metamorphs are a bright, nearly unpatterned green when they move out into the grassy areas surrounding the breeding ponds. Frog tadpoles are notoriously hard to tell apart, but Gray Treefrog tadpoles often have an orange to reddish wash on their tail.

Wood Frog
Lithobates sylvatica

Type specimen described by John Eatton LeConte in 1825, collected from the vicinity of New York City

Head–body length: metamorph = 0.6 inches (1.4 cm) to adult = 3.3 inches (8.3 cm)

The Wood Frog, formerly *Rana sylvatica*, is found throughout the region with the exception of southeastern and south-central Virginia. Its range extends farther north than any other amphibian or reptile in North America, reaching above the Arctic Circle in Alaska and the Yukon. A basically forest-dwelling species, it survives in the far North in tundra habitat where trees do not grow.

This is the frog with the Zorro mask. A dark band extends from its nose across its eye, ending in a triangular point behind the tympanum, or "ear." The general background color is light tan to dark brown or even pinkish. Dark cross-bands on the thigh, lower leg, and foot of the hind legs line up when the frog is tucked in a crouching position. The sides of its upper lip are lined with yellow, off-white, or tan. Females in a population tend to be lighter colored and larger than the males.

In most areas the Wood Frog is the first frog to begin breeding choruses and certainly the first to end them, being an explosive breeder that takes just a few days to a week to complete its late-winter mating rituals. A study comparing calling dates of Wood Frogs, Spring Peepers, Gray Treefrogs, and Bullfrogs in the vicinity of Ithaca, New York, from the 1990s demonstrated that all of these species are calling 10 to 13 days earlier than they did 100 years earlier. During this period the temperature rose about 4.1°F (2.3°C), a potential sign of the impact from global climate change.

Breeding occurs almost exclusively in fish-free wetlands or pools. Even in larger shallow ponds or wetlands with lots of open water, the breeding adults tend to congregate in select small areas to lay their eggs. Numerous frogs in amplexus can be observed during the day and at night during the height of their breeding activity. The result is large mats of egg masses touching each other, making it impossible to get an accurate count. The mats can extend for more than 33 square feet (3 m²), representing hundreds of mated pairs and tens of thousands of eggs. Such large congregations of eggs and recently hatched larvae are a readily available source of protein for Red-spotted Newts, Spotted Turtles, leeches, and aquatic insects.

The short stay at the breeding pond is followed by a long excursion in nearby woodlands. Wood Frogs have one of the longest activity periods, being one of the last anurans to retreat to an overwintering site. They grow and mature quickly, rarely surviving more than three or four years. In a mature forest with a healthy layer of leaf litter and freshly fallen new leaves, selection of an overwintering site is not difficult. Fortunately, the Wood Frog has a special adaptation that helps it withstand the long, cold winters. They simply bury beneath the leaves in an area where they won't desiccate. But they do not need to retreat to below the frost line. Instead, they produce glucose, an antifreeze-type chemical, and urea, enabling them to freeze without causing lethal cell damage. Ice does form but in the blood and outside, rather than inside, the cells.

Spring Peeper
Pseudacris crucifer

Type specimen described by Prince Alexander Philipp Maximilian zu Wied-Neuwied in 1838, collected from Leavenworth County, Kansas

Head–body length: metamorph = 0.8 inches (2.0 cm) to adult = 1.5 inches (3.7 cm)

A small frog with a piercing voice, the Spring Peeper is indeed a sign of spring to come. Its choruses begin in late winter and continue to early June. At first, just a few isolated males call back and forth, but when the time and temperature are right, the calls come so fast and furious that the sound is deafening when heard from the edge of the pond. It is next to impossible to determine from what direction Spring Peepers are calling. Those closest to the intruding observer become silent, with the cone of silence moving with the observer as she wades through the wetland. During the height of the breeding season, choruses can be heard during the day, but they reach peak intensity after dark. The high notes of the Spring Peeper are inversely related to its body size when compared to the low bass notes of the Bullfrog with its huge body, but the length of the vocal chords are responsible for the difference in the calls. Smaller male Spring Peepers produce higher-pitched calls than larger Spring Peepers and Bullfrogs, just as the shorter strings on a musical instrument produce higher notes than the longer strings.

It is difficult to determine a location with the eyes—that is, a search image—when the ears keep telling the brain that the target species is now in a different location. But after several frustrating minutes, the search image clicks into focus, and these small frogs do become more obvious. During the height of their breeding choruses, you will see them tucked away at the base of the vegetation on a mossy or sedge hummock, often with short, overhanging shrubs holding the tussock together. You might also see them calling while floating nearly motionless on the surface of the water. Those individuals with the most alluring songs may already be in amplexus, with the receptive female at or below the surface of the water. But you won't see the egg masses so typical of the Wood Frog or the ambystomid salamanders. The Spring Peeper lays its eggs singly or in small clusters attached to submerged vegetation, with each tiny female laying 200 to 1,200 eggs!

Although so obviously associated with the wetlands through its intensive bouts of chorusing, the Spring Peeper actually spends most of its year in the surrounding upland forest, foraging on the surface in summer and hibernating below the leaf litter in winter. While hiking in mature woodland habitat in the summer on days following rainy nights, it is not uncommon to detect a flash of movement in the dry leaves that may easily be dismissed as an insect. On closer examination, it may well be a Spring Peeper camouflaged against the dead leaves in a light tan to pinkish-tan or darker brown color with the signature "X" inscribed on its back. At these times, when the air is damp or a light rain is falling, the Spring Peeper may surprise you by issuing a few notes of its advertisement call. The Spring Peeper is the only frog of the North whose call may be heard every month of the year, even during a brief thaw in January or February.

Upland Chorus Frog
Pseudacris feriarum

Type specimen described by Spencer Fullerton Baird in 1854, collected from Carlisle, Pennsylvania

Head–body length: metamorph = 0.5 inches (1.2 cm) to adult = 1.5 inches (3.8 cm)

The Upland Chorus Frog was previously considered to be a subspecies of the Western Chorus Frog (*Pseudacris triseriata*). In appearance, it differs from the western form by having three longitudinal stripes that are thinner and often appear as dashed lines or series of spots. In the Northeast, many miles separate the two species, with the Upland Chorus Frog being found throughout most of Virginia and Maryland, eastern West Virginia, southeastern Pennsylvania, and northern New Jersey. The Western Chorus Frog is found in westernmost West Virginia, Pennsylvania, and New York. A third species, the Boreal Chorus Frog (*P. maculatum*), was recently separated from the Western Chorus Frog and is found along the New York–Canadian border area of the St. Lawrence River, previously reaching into Vermont along Lake Champlain. In southern New Jersey and the Delmarva Peninsula, the New Jersey Chorus Frog (*P. kalmi*) is found, a species that is also found in Pennsylvania, where it is endangered.

Chorus Frogs are placed in the Treefrog Family, but they are not arboreal, being poor climbers. Most often they are found in seasonal wetlands during early spring breeding by their chorus, a not very musical, clicking trill. They are opportunistic in their choice of breeding ponds and can successfully reproduce in shallow pools in old fields, roadside ditches, and water-filled tire ruts left by heavy vehicles driving through wet meadows.

As with most amphibians, their journey from egg to larva to land is a difficult one, each stage presenting its own challenges. Perhaps the biggest challenge is transforming from an aquatic organism to a terrestrial one, a process known as metamorphosis. This transformation requires much more than simply leaving the water. Changes in external morphology coincide with changes in internal anatomy, physiology, and the ecological niche into which the species fits.

In the aquatic form, there are a few significant differences between salamander larvae and frog larvae, also commonly called tadpoles. Most frog tadpoles are grazers, feeding on algae, fungi, and surface layers of other vegetation with their rasping-type teeth. Most salamander larvae are carnivores. As a salamander larva matures, it develops four legs of approximately equal size at about the same time. Frog tadpoles first develop hind legs. On approaching the time for metamorphosis, the front legs burst through the skin and emerge as reasonably well-developed limbs already equipped with toes. To move about on land, a frog needs four legs. But the tail becomes a hindrance, and as these changes progress, the tail is reabsorbed into the body. The aquatic tadpole has gills to absorb oxygen from the water column and rid the body of carbon dioxide. To move out on land, it must have lungs to take in atmospheric oxygen and dispel carbon dioxide. The frog's mouth also changes from the small, circular rasping oral disc to a large mouth with a long tongue to snatch up prey. Its skin thickens, eyelids develop, and a host of other changes take place in preparation for its new terrestrial lifestyle. It is not easy to leave the water.

Eastern Spadefoot
Scaphiopus holbrookii

Type specimen described by Richard Harlan in 1835, collected from the vicinity of
Charleston, South Carolina

Head–body length: metamorph = 0.4 inches (0.9 cm) to adult = 2.9 inches (7.3 cm)

Endangered in Connecticut and Pennsylvania

Threatened in Massachusetts and Rhode Island

The Eastern Spadefoot is an explosive breeder with the shortest tadpole period of all northeastern anurans. Unlike other frog species, its breeding events cannot be predicted by checking a calendar. It may even go a few years without breeding at all. What it is waiting for is rain. And not just a little rain—lots of it. Ground-soaking rain. Torrential rain. An inch of rain will not do. Even two or three inches of rain might not bring Eastern Spadefoots to their breeding pools if the previous period had left a large moisture deficit. Rain that lasts several days, hurricane-intensity rainfalls, or nor'easters will set the stage for a breeding chorus of Eastern Spadefoots. These breeding events could be any time from March to September.

The Eastern Spadefoot has the most nasal voice of all northeastern frogs. It has been variously described as a grunting sound, the call of a young crow, or a person gagging. Regardless of how you interpret the sound, it is unlike any other frog call in the Northeast and will illicit notice. When searching for new Spadefoot breeding areas at night, someone in our group will inevitably stop and say, "Listen—did you hear that? I heard something; not sure what it was." After following the call for several hundred yards, the sound turns out to be the call of a Spadefoot starting a chorus that may end up as just a small handful of animals or a breeding chorus of 500 or more adults. The call comes in short bursts and is repeated at short intervals. The chorus may begin an hour or so after dark and increases in intensity as more males arrive at the breeding pool, perhaps reaching a crescendo around midnight. My friend Scott Smith, from the Delmarva Peninsula in Maryland, said at the height of a hurricane he could hear the chorus of Eastern Spadefoots from inside his house with the windows closed. But the breeding frenzy is short lived. Spadefoot choruses last one night, maybe two, and then the adults retreat to their upland burrows that they dig with the spade on their back feet that they use to back into their excavation.

The breeding ponds that Eastern Spadefoots select are sometimes not terribly obvious to humans. A slight depression in the forest or field from just a few inches to three feet deep will serve. Almost exclusively formed in sandy soil, with no sign of the wetland vegetation that is present in other types of seasonal wetlands, these pools may last just three or four weeks. That is sufficient time for the eggs to be laid, as larvae emerge within one or two days and transform into the terrestrial stage in as little as two weeks. Spadefoot metamorphs and adults can be found foraging at night when the ground is damp. In exposed areas of sand, small depressions reveal where Spadefoots emerged from their burrows. Two yellow lines on their back form a misshapen hourglass pattern on a generally brown back. Spadefoots are quite toxic, not just to most predators that attempt to eat them, but to humans as well.

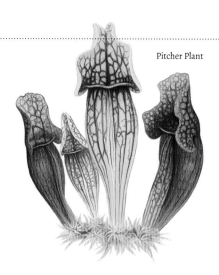

Pitcher Plant

5 Bogs

ONTRARY TO popular opinion, not all shallow, mucky wetlands are bogs. From a botanical perspective, bogs, fens, and wet meadows each have their own unique set of characteristics. Also contrary to popular belief, wet meadows and fens form not just in flat areas but also in areas with significant slopes. These habitats support nearly one-third (52 species) of the herp species in the Northeast.

Bogs are highly acidic wetlands; that is, using the standard scale to measure acidity versus alkalinity, bogs fall out with a pH of 5.0 or less on a scale where a pH of 7.0 is neutral. A low pH means there are fewer available nutrients, so that plant species adapted to living in an acidic environment are the ones that thrive. Carnivorous plants, which can derive nutrients from the insects they capture, such as Pitcher Plants (*Sarracenia purpurea*) and Sundews (*Drosera* spp.), are found in these systems. *Sphagnum* mosses, commonly referred to as peat mosses, are usually dominant in a bog, and the peaty deposits underlying a bog mat are mainly composed of dead *Sphagnum*. Bog habitats tend to have a lower diversity of amphibians and reptiles than are found in the more nutrient-rich fens and wet meadows. Larval survival of the Spotted Salamander decreases in wetlands with a pH of less than 6.0, but Blue-spotted Salamanders do breed in ponds as acidic as pH 4.5, and Wood Frogs can successfully breed in ponds with a pH of 4.0.

Bogs primarily obtain nutrients and water by precipitation, with little or no influence from groundwater sources such as feeder streams, springs, or seeps. Rain and snow carry few nutrients and in recent decades have become more acidic due to anthropogenic sources of air pollution. Bogs are formed in depressions left by retreating glaciers and generally have no inlet or outlet. The classic kettle-hole bog is formed in a deep circular depression with concentric rings of vegetation surrounding open water in the center. At the edge of the open water is a floating mat of *Sphagnum* topped with knee-high shrubs and sedges. This floating vegetation, called a quaking bog mat because it will undulate under the weight of a human or other large mammal, may hide water of a considerable depth. Toward the shore, the vegetation on the mat becomes progressively taller. In a mature kettle-hole bog, taller trees such as Tamarack (*Larix laricina*) and Black Spruce (*Picea nigra*) connect the bog to the upland. Commercial mining of sphagnum peat from bogs results in the loss of all surface vegetation and associated wildlife species from these areas.

Fens are wetlands fed by subsurface groundwater and spring seeps. These water sources contain greater amounts of dissolved minerals than rainwater, and hence the pH is not as acidic as in bogs. The ground flora is dominated by graminoid or grass-like species such as grasses, sedges, bulrushes, and reeds. The presence of tussock-forming sedges generally initiates the formation of hummock-and-hollow microtopography within the fen. In the more alkaline fens with pH from near neutral to slightly above pH 8.0, Grass-of-Parnassus (*Parnassia glauca*) and Shrubby Cinquefoil (*Dasiphora fruticosa*) may thrive. *Sphagnum* may be present, usually as raised hummocks around the stems of short shrubs. Fens are generally open-canopied wetlands, but a lack of grazers and browsers may result in the development of a partial tall shrub or tree canopy, making the system less suitable for species that prefer early successional habitats in which to nest and forage, such as Bog and Spotted Turtles.

Wet meadows are wetland systems dominated by graminoid species on poorly drained soils with a water table that is close to the surface. Both rainwater and surface water feed wet meadows, which often form in areas of abandoned beaver impoundments and historically were prime grazing areas for Elk and Bison. Lacking in trees, wet meadows were easier to convert into agricultural uses than the surrounding forests. While operated as pastureland, they still provided wetland benefits to many herp species, but once drained for crops, the value for wildlife diminished greatly.

Black dirt and muckland farming derive their names from the dark, extremely fertile, fine-textured organic soils that developed in 14,000-year-old shallow glacial lake bottoms. Muckland farming is ideal for producing root crops such as onions, carrots, radishes, and potatoes but it is also good for crops such as lettuce and tomatoes. The Black Dirt Region of the Walkill River floodplain region of northern New Jersey and southeastern New York, containing 26,000 acres (10,400 hectares) of muckland, is the largest such area in the United States with the exception of the Florida Everglades. Other large areas of muckland converted to agriculture are scattered across the Lake Plain region of New York just south of Lake Ontario. Where natural vegetation exists on muckland, the sites often support a rich diversity of herps and other wildlife.

Four-toed Salamander
Hemidactylium scutatum

Type specimen described by Hermann Schlegel in 1838, collected from
Nashville, Tennessee

Total length: hatchling = 0.4 inches (1.1 cm) to adult = 4.0 inches (10.2 cm)

The Four-toed Salamander is the smallest salamander in the Northeast and easy to overlook. Four-toed Salamanders are associated with wetlands where *Sphagnum* moss is prevalent, but because they spend most of the year and overwinter in uplands, they may be found at considerable distances from the boggy habitat. One of the best times to search for them is on those same rainy late-winter or early-spring nights when the ambystomid salamanders, Wood Frogs, and Spring Peepers are migrating to their breeding ponds. At first glance, people participating in road-crossing surveys might mistake a Four-toed Salamander for a Red-backed Salamander, but the slightly stockier body, a blunt and squared-off nose, and a noticeable constriction at the base of the tail should cause a second glance. If flipped over, an alabaster-white belly, boldly marked with jet-black dots is revealed. Note also that, as its name implies, it has only four toes on each hind foot as well as on its front feet. Most other salamanders have five toes on their hind feet.

The Four-toed Salamander has aquatic larvae but does not lay its eggs in water. Instead, it lays its eggs deep in *Sphagnum* moss. *Sphagnum* moss is ideal because its cellular structure allows it to absorb large quantities of water to sustain the plant during drier periods, which of course is a perfect place for a salamander to hide its eggs so they do not desiccate while the embryo is developing. But not every clump of *Sphagnum* serves as a possible nest site. Ideally, the *Sphagnum* grows in an elevated tussock above the water surface, and it is usually supported by the root crown and stems of low shrubs so that the edge of the moss overhangs the water surface. When the eggs hatch, the tiny larvae work their way to the end of a moss shoot and drop into the water, where they continue their development.

The constriction at the base of the tail is not just for decoration. Many salamanders survive encounters with predators after losing a section of their tail, but the Four-toed Salamander has evolved a different strategy. Its tail will break off easily and cleanly at the constriction, confounding a predator that turns its focus to the tail, which continues to twitch violently while the remainder of the salamander scurries away.

During most of the active season, Four-toed Salamanders are found under logs, stones, or leaf litter in the woodlands. A second peak in observations occurs in late September or early October, when adults and young of the year are moving back toward overwintering sites. At that time the experienced observer may find small groups of them congregating within a small area.

Northern Red Salamander
Pseudotriton ruber ruber

Type specimen described by Pierre André Latreille in 1801, collected from the vicinity of Philadelphia, Pennsylvania

Total length: metamorph = 3.0 inches (7.5 cm) to adult = 7.1 inches (18.1 cm)

The Northern Red Salamander is the Scarlet Tanager of the herp world: candy-apple red with small jet-black spots abundantly and randomly scattered across it back, sides, and chin. Notice the yellow eye, which contrasts with the vibrant red body color. Maturing larvae are usually a brownish color, but shortly before transforming they become a salmon-pink color that they retain as juveniles. Really ancient specimens, possibly more than 20 years old, turn quite dark, purplish black with the spots blending into the background color, and barely a hint of the more youthful brilliant red. The similarly marked Mud Salamander has a brown eye and an unspotted chin.

The Northern Red Salamander reaches it northernmost point in the Hudson River Valley of New York just south of the Mohawk River–Erie Canal. The Mohawk River seems to have created a barrier to northern migration of several species of amphibians and reptiles such as the Northern Red Salamander, a species found in well-oxygenated seeps, but also includes the Northern Slimy Salamander, a completely terrestrial species, and the Eastern Wormsnake, a fossorial species. Yet the Mohawk River does not form a barrier to other herp species in those same three guilds, so the limits to their northward expansion following the last glacial period must be based on other yet-to-be identified life history traits.

The Northern Red Salamander is a creature of cool-spring seeps and small flowing rivulets in areas of mixed-hardwood forests. The adults spend their active season in fen or wet meadow habitat that is fed by the spring seeps, returning to the spring seeps to breed and overwinter. The temperature of these seeps varies with the northern latitude and elevation, reaching a minimum at the northern limits of its range. I have found adult Northern Red Salamanders at night on the surface of a seep outflow in mid-January when the air temp was 25.0°F (−3.9°C) and the water was 42.0°F (5.6°C), with active larval Red Salamanders under nearby rocks. Presumably, even in the depth of winter, they are searching for prey.

The female deposits her eggs on the undersides of rocks deep in underground caves or spring seeps in early fall. Eggs hatch in late December or early January. The larvae take nearly 4 years to reach maturity and transform into the adult form. These metamorphs reach sexual maturity about 1.5 years after transforming, thereby taking about 6 years to progress from newly deposited egg to reproductive adult.

The adult Northern Red Salamander is a voracious predator, feeding on all manner of invertebrates and small salamanders, including both aquatic and terrestrial species such as Red-backed and Slimy Salamanders. As with the Red Eft, Northern Red Salamanders have toxic chemicals in their skin. The bright red color is a warning to some potential predators that they may be quite distasteful and should be left alone.

Pickerel Frog
Lithobates palustris

Type specimen described by John Eatton LeConte in 1825, collected from the vicinity of Philadelphia, Pennsylvania

Head–body length: metamorph = 0.7 inches (1.9 cm) to adult = 3.4 inches (8.7 cm)

It's the spotted frog with the square spots. If you can get a good look before it leaps, the shape of the spots will confirm a Pickerel Frog. But it can be confused with the Northern, Southern, or newly described Atlantic Coastal Leopard Frog (*Lithobates kauffeldii*) when the spots are not quite as square as they should be. The Pickerel Frog and all Leopard Frogs have the prominent bronze-colored dorsolateral ridges or folds that run from behind the eye almost to the groin. The general body color is usually more of a light brown, while Leopard Frogs frequently have greenish hues. Young Pickerel Frogs may have a metallic hue to their basic color. With legs extended, the bright yellow wash above the hips and extending to the thighs and lower belly distinguishes the Pickerel Frog from the Leopard Frog. The frog that "peed its pants," as an old professor of mine used to say, the "pee" will help you remember Pickerel as the frog with the yellow wash.

The yellow may also be a warning to potential predators that it is a distasteful species. Leave it alone and eat a Leopard Frog instead. Pickerel Frogs secrete a substance from their skin that is toxic to some predators and can be irritating to people who handle it and transfer the secretions to mucous membranes or scratches. The Pickerel Frog is the only frog in the family Ranidae in the United States or Canada that is poisonous. Most snakes and mammals prefer to leave the Pickerel Frog alone, but, as with most similar defenses, some predators have evolved immunity to the toxin. The Pickerel Frog was most likely named by early fishermen because of its use as bait for its namesake, the Pickerel (*Esox niger*), although Pickerels prefer to eat minnows instead.

The Pickerel Frog has a rather short breeding season of just one or two weeks. Its call is a not-so-loud snoring sound that lasts only a few seconds. Even in full chorus the sound does not travel long distances. The Pickerel Frog makes it call from a pair of vocal sacks situated on either side of its head, looking much like huge cheek pouches. Somewhat strangely, it also makes its call while being fully submerged rather than above the water surface like most anurans; hence the sound does not travel very far.

The Pickerel Frog's range includes the entire Northeast region. After breeding season is over, the Pickerel Frog is usually encountered in upland areas, ranging from the shoreline of ponds and streams to far afield in grasslands and meadows. As winter approaches, it retreats to an aquatic habitat with well-oxygenated flowing water. In mid-January, with the surrounding landscape buried in snow and ice, I have found large numbers of them hiding under rocks, not quite completely dormant, in spring seeps where the water temperature never falls below 42.0°F (5.6°C).

Western Chorus Frog
Pseudacris triseriata

Type specimen described by Prince Alexander Philipp Maximilian zu Weid-Neuwied in 1838, collected from Posey County, Indiana

Head–body length: metamorph = 0.4 inches (0.9 cm) to adult = 1.5 inches (3.9 cm)

I was camping with my family on the north shore of Lake Superior in Minnesota. As it got dark, we began to hear the distinct call of the Western Chorus Frog, which is similar to the sound made by running your thumb along the teeth of a comb. The call is much more muted than that of its cousin the Spring Peeper, and it is more difficult to actually see a Western Chorus Frog. My two young daughters and I each grabbed a flashlight and went in search of the frog. We found a small isolated pool, about the size of a card table, perched on top of bedrock that rose up from the lakeshore. A small ring of vegetation—mostly graminoids, mosses, and a small shrub—bordered our pool. *This will be easy*, I thought. We quietly circled the pool, waited until we heard distinct chirping so we could tell where to point the lights, and then on the count of three we all turned on our lights. Nothing. For a couple minutes we strained our eyes looking for this one frog. Did he (only the males make these calls) escape as soon as we turned on the lights? So we turned off the lights. We waited a few minutes and the call resumed, apparently at the exact spot where we had thought he was. Lights on. Nothing. Lights out. Calling resumed. This scenario went on for 45 minutes until finally my younger daughter, perhaps 10 at the time but with eyes keener than mine, found our quarry.

It is not just us. This frog is easy to hear but hard to see in the field. A quick analysis of the New York Amphibian and Reptile Atlas Project data on frog vocalizations indicates that this frog is heard more often than seen, at a rate higher than any other species of frog in New York. Fully 88% of all Western Chorus Frog reports were based on hearing them call rather than actually seeing one. The elusive Spring Peeper was reported only 77% of the time based on calling, and the American Toad, a very vocal species, was heard rather than seen only 19% of the time.

Perhaps this is why for years it was thought that there was only one species of chorus frog in the northern portion of Pennsylvania, New York, and Vermont. Recent genetic analysis tells us we have two species. The Western Chorus Frog is found along the south shore of Lake Erie and Lake Ontario as far east as Oswego County, New York. Along the east shore of Lake Ontario, north along the St. Lawrence River, then dipping south along both the east and west shores of Lake Champlain, is where we find the Boreal Chorus Frog (*Pseudacris maculata*). Used to find, that is. Recent surveys indicate that this species is no longer found along Lake Champlain. It is listed as endangered in Vermont, but it may be extirpated. And the Western Chorus Frog is now missing along large areas bordering Lake Erie in western New York and Pennsylvania.

Spotted Turtle
Clemmys guttata

Type specimen described by Johann Gottlob Schneider in 1792, collected near Philadelphia, Pennsylvania

Carapace length: hatchling = 1.0 inches (2.5 cm) to adult = 5.0 inches (12.7 cm)

Endangered in Vermont

Threatened in Maine and New Hampshire

A specimen with a plain black shell with one or more circular yellow spots on each scute of the carapace can only be a Spotted Turtle. The yellow spotting continues onto its head, neck, and front side of its legs. Hatchlings typically have only one spot per scute, but more spots usually appear added as the turtle matures. The plastron is yellow with dark blotches toward the outer edges. The yellow spots on the carapace may become obscure from tannins in the wetland. A larger yellow-orange marking on each side of the head and neck may cause an observer to think briefly that it is a Bog Turtle, but the spots on the carapace will be the deciding factor. Typical of a number of other turtles, the male has a longer tail than the female, and its plastron is concave compared to the female's nearly flat plastron.

The Spotted Turtle uses a variety of wetland habitat types and will frequently travel a half mile (0.8 km) or more overland or along forested stream channels to reach these various components of its home range. If a wetland system is large enough, a Spotted Turtle can overwinter, mate, nest, and forage without ever leaving the confines of the wetland. In smaller systems, the Spotted Turtle may overwinter in the wetland but leave it early in the spring, traveling through upland areas to reach vernal pools. At the pools, the Spotted Turtle will feed heavily on Wood Frog and ambystomid salamander egg masses and larvae.

Where the microhabitat is suitable, they hibernate communally with other Spotted Turtles and Bog Turtles. They emerge from hibernation one to two weeks earlier than Bog Turtles and can often be seen basking in early spring when water temperatures are below 55.0°F (12.8°C) and air temperatures are above 55.0°F (12.8°C), or when bright sunshine warms favored basking sites. A small species, the Spotted Turtle lays a small clutch of two to seven eggs. In a few cases, Spotted Turtles have been known to nest twice in one season. Their nests are formed in open-canopy areas on the top of sedge tussocks, in mounds of sphagnum moss, or in loamy soils where late hatching may result in the young overwintering in the nest.

These small turtles are slow to mature, taking 10 to 15 years after hatching to reach sexual maturity. In the past they have been prized as pets for their small size, ease of captive maintenance, and, most importantly, attractive shells. In 1905, Spotted Turtles were reported to be as common as Painted Turtles in the vicinity of New York City, but that is certainly not true today. Because of unregulated collecting of the past, illegal collecting today, and loss of wetland habitat, the Spotted Turtle's future is not as rosy as conservationists would hope.

Bog Turtle
Glyptemys muhlenbergii

Type specimen described by Johann David Schoepff in 1801, collected near Lancaster, Pennsylvania

Carapace length: hatchling = 0.8 inches (2.0 cm) to adult = 4.5 inches (11.4 cm)

Federally threatened

Endangered in Connecticut, Delaware, Massachusetts, New Jersey, New York, Pennsylvania, and Virginia

Threatened in Maryland

Extirpated from the District of Columbia

If you would like to see a Bog Turtle, try the gardens at Muhlenberg College in Allentown, Pennsylvania. The Bog Turtle, or Muhlenberg Turtle as it was originally called, is the unofficial mascot of this college named for the father of the clergyman Reverend Gotthilf Henry Ernest Muhlenberg, who first found one in a wetland near Lancaster. The turtle in the college garden is a bronze sculpture, so it is easily found. Live, wild, and free-roaming Bog Turtles are much harder to find. It helps to be a botanist or at least someone who is intimately familiar with wetland vegetation.

Reverend G. H. E. Muhlenberg was a botanist who specialized in grasses and sedges, primarily wetland species. I can easily envision Reverend Muhlenberg accidently discovering his namesake turtle while intently searching for a new species of wetland plant, because this is exactly how my botanist friend Bob Zaremba found two new Bog Turtle sites in Upstate New York. My kinship to the reverend is twofold: I grew up just 2.3 miles (4.8 km) down the road from where he was born in Trappe, Pennsylvania, in 1753, and my early training was as a botanist. For me, the search for Bog Turtles is enhanced by the floristic beauty and diversity of the wetlands where these turtles live. A day spent in one of these sites is a good day even when I fail to find a Muhlenberg Turtle.

Bog Turtles are creatures of wetlands with shallow water where they will spend their entire life hibernating, mating, nesting, and feeding. They have been described as a species that prefers their feet wet and their upper shells dry, but in truth they are often found completely buried in the deep, soft, peaty muck, which is characteristic of their preferred habitats. Typical searches often involve muddling—crawling through the wetland on hands and knees, and reaching under sedge tussocks or down into abandoned muskrat or other small mammal burrows. Bog Turtles are now a protected species in all states where they occur and are listed by the US Fish and Wildlife Service as threatened in all states from Maryland north, so these types of surveys require a permit from the US Fish and Wildlife Service and the respective state.

Most people searching for Bog Turtles focus on graminoid fens—groundwater-fed wetlands that are dominated by grasses, sedges, and rushes—that are usually calcium-rich, high-pH systems. A surprising feature of these fens to many observers is that they are not formed in flat valleys but rather on gently sloping hillsides. But these are not bogs in the botanical sense, so the name Bog Turtle seems out of place, and the name Fen Turtle is perhaps more fitting. The term "bog" brings to mind floating mats of vegetation that can be treacherous to walk on. Bogs are fed by rainwater rather than groundwater, resulting in an acidic condition where low pH encourages cranberries, sphagnum mosses, Pitcher Plants, and Sundew to thrive. In a "young" bog, the center is open water encircled by the floating mat. The turtles found at these sites truly are Bog Turtles. A canoe is a handy device to survey these sites. I know of only four such sites, all in Upstate New York.

The wetlands where Bog Turtles are found can range from sites not much bigger than the size of a small house to wetlands that are several thousand acres. Finding a 4.0-inch (10.2-cm) turtle in a wetland the size of a house may take an hour or two per turtle. Finding a 4.0-inch (10.2-cm) turtle in a 1,000-acre (405-hectare) wetland requires a trained eye to detect the subtle differences in vegetation and hydrology, followed by many hours of searching.

Massasauga
Sistrurus catenatus

Type specimen described by Constantine Samuel Rafinesque-Schmaltz in 1818, collected from prairies of the Upper Missouri

Total length: at birth = 7.5 inches (19.1 cm) to adult = 39.5 inches (100.3 cm)

Endangered in New York and Pennsylvania

The Massasauga is the smallest and rarest venomous snake in the Northeast. Although widespread in the Midwest, in the Northeast it is restricted to a few remaining sites in northwestern Pennsylvania and two sites in New York. The Massasauga is the only venomous snake in Canada, where it is restricted to the Great Lakes area of Ontario Province.

The Massasauga has a gray body color attractively highlighted with a row of large brown to reddish-brown hourglass or roundish-shaped blotches the length of its back with three rows of staggered blotches per side, the blotches in each row smaller than the one above. Nine large scales on the top of the head of the Massasauga separate it and other species of *Sistrurus* from the Timber Rattlesnake and other members of the genus *Crotalus* that have numerous small scales on the crown with just a few larger scales bordering the crown.

Unique among snakes, New World rattlesnakes are the only group that have evolved a "rattle," that is, a loose collection of horny segments at the end of the tail. A new segment forms each time the snake sheds. When the Massasauga shakes its tail, the sound is not unlike the buzz of ground-nesting yellow jackets. When you hear that sound you usually have time to stop, look around, and step back if indeed the sound comes from a rattler. If it is from yellow jackets, too late—you will probably be stung.

The name Massasauga is from the Ojibwa or Chippewa language and is believed to mean "great river-mouth," alluding to the wetland habitat where the snakes were found. European settlers referred to the snake as simply the "swamp rattler" because of its predilection for wetlands. Historically, the range of the Massasauga coincided closely with that of the tallgrass prairie in the Midwest, which had fingers reaching into western Pennsylvania and New York. The loss of the tallgrass prairie, and draining of the shallow wetlands for agriculture where the snake made its home, led in turn to the decline of the Massasauga over much of its range.

Although the wetland is where they overwinter and where females give birth to young, Massasaugas do leave the wetlands in summer to forage in the surrounding grasslands, agricultural lands, and forests for field and jumping mice, voles, and shrews. Because its primary habitats are early successional wetlands and grasslands in the Northeast, the species is threatened by abandoned agricultural lands reverting to woodlands and wet meadows, bogs, and fens reverting to closed-canopy swamps. Both New York and Pennsylvania are working to restore these habitats through selective cutting, clearing, and prescribed burning.

Eastern Ribbonsnake
Thamnophis sauritus sauritus

Type specimen described by Carolus Linnaeus in 1766, collected from the vicinity of
Berkeley County, South Carolina

Total length: at birth = 7.2 inches (18.4 cm) to adult = 40.1 inches (101.8 cm)

Two subspecies of the ribbonsnake are found in our area, the Eastern Ribbonsnake and the Northern Ribbonsnake (*Thamnophis sauritus septentrionalis*). The Northern Ribbonsnake was not described as a separate subspecies until 1963, based on a specimen collected in Michigan Hollow, near Ithaca, New York, a favorite stomping ground for professors and students from Cornell University.

The two subspecies of ribbonsnakes are often confused with the Common Gartersnake. Ribbonsnakes do have a few distinct features that separate them from other members of the genus *Thamnophis*. They are more slender and have much longer tails. Of course, when an animal doesn't have hind legs, deciding at a glance where the body stops and the tail begins might not be obvious to the casual observer. If you were to examine a ribbonsnake's ventral surface, you would find that the body ends and the tail begins at the cloacal opening, a slit that runs perpendicular to the length of the snake. Ribbonsnakes have tails that are fully one-third of the total body length, whereas other *Thamnophis* species have tails that are less than one-quarter of the total length.

The Eastern Ribbonsnake has a light to dark brown body with three sharply defined yellow stripes, one vertebral stripe and one on each side encompassing the third and fourth scale rows. The Northern Ribbonsnake is even more impressively colored, with a dark, almost velvety black body that accents the bright yellow stripes to an even greater degree. Ribbonsnakes have the general appearance of a gartersnake that has been polished to give it a lustrous shine. Some specimens may have a hint of the checkerboard pattern between the stripes that is more frequently seen on the Common Gartersnake.

With their primary foods being amphibians and fish, ribbonsnakes are denizens of many types of aquatic systems, from wet meadows and fens to vernal pools to small streams, rivers, and lakeshores. In the spring, they emerge from hibernation about the same time that the ambystomid salamanders and Wood Frogs migrate to their breeding ponds. Ribbonsnakes prey on both larval and adult salamanders, tadpoles and adult frogs, and fish, selecting food items that are appropriate to their size as they grow from juveniles to mature adults.

The ribbonsnakes' color pattern is striking and obvious when in the open, but it serves them well when they are fleeing or attempting to hide in the cover and shadow of grasses, sedges, and emergent wetland vegetation that dominate their habitats. They move quite quickly and are difficult to catch, but they are not likely to bite when handled. Ribbonsnakes are moderately accomplished climbers and can be found in shrubs and bushes as high as 5.0 to 6.0 feet (1.5 to 1.8 m) above the water surface.

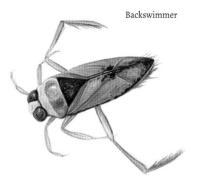

6 Headwaters

F LOWING WITHIN 150 feet (46 m) of our camp in the southern Adirondacks is a small, ephemeral stream arising from a seep about 400 feet (122 m) upslope of our log cabin. In late spring the flow stops, and by early summer in most years even the few small remaining surface pools have disappeared. The streambed is mostly coarse sand mixed with sediment derived from decaying leaves and other vegetation trapped by tree roots and downed logs, and underlain by glacial cobbles to an unknown depth. In the spring, before we dug a well, the stream was our source of drinking water until one day I found a first-year larval Northern Two-lined Salamander swimming in my glass. Because I had scooped the water from an area that is bone dry in the summer, I was amazed that a salamander species requiring at least two years to transform had somehow successfully reproduced in our stream. But I was even more surprised years later when I found a third-year larval Northern Spring Salamander in that same ephemeral stream! And that was after having explored the stream and surrounding forest for 17 years. Those larval salamanders, and the adults, must have lived lives that were almost completely subterranean, surviving in the interstitial spaces between the glacial rocks. Since then, I have found other ephemeral streams that support populations of Two-lined and Dusky Salamanders. Perhaps it is not unusual or unsuspected. The species of salamanders that can make their home in caves that humans visit should also be able to make their homes in caves that only salamander-sized critters can visit.

Headwater stream habitats are in many ways unique. Generally a person can step across one of these streams or hop across in just a few short leaps. Often they are steep-gradient streams, not navigable waters by any US Army Corps of Engineers definition. But they are a favored habitat for a suite of salamanders. And, as I have more recently discovered, spring seeps are overwintering habitat for several species of frogs in addition to the salamanders. Springs and seeps are groundwater discharge points, making them cooler in the summer but warmer in the winter than the surrounding habitat. With a nearly constant temperature approximating the average yearly temperature of the area's groundwater, water from spring seeps ranges in the Northeast from as warm as 62.0°F (16.7°C) in southern Virginia to as cool as 42.0°F (5.6°C) in northern New York and New England. Cool water is richer in dissolved oxygen than warm water, a definite plus to stream salamander larvae, which have smaller external gills than pond salamanders.

Headwater streams usually lack fish that prey on amphibian eggs, larvae, and adults. As the stream gradient lessens downstream, fish become more abundant, and salamander abundance and diversity drop. Rocks and woody debris within the streams provide structure for salamanders, frogs, and their larvae and act as attachment points for the eggs of some species. Springtails (order Collembola), most of which are less than one-quarter inch (6 mm) long, are seasonally abundant around the discharge point of some seeps, providing food for larval and adult salamanders alike.

The integrity of small steams relies on the surrounding habitat to provide the best conditions for amphibians. Under natural conditions, this means having a forest canopy bordering the stream that helps maintain moderate temperatures and reduce siltation. Until recently, common logging and agricultural practices involved clearing the land to the edge of the stream. Forests were clear-cut to the water's edge and logs were skidded across the streams, greatly increasing stream bank erosion. Streams flowed through pastures, allowing livestock to graze on the bordering grasses while standing in flowing water. Both of these practices have been curtailed in many areas, but impacts to these small streams will take years to heal.

Northern Dusky Salamander
Desmognathus fuscus

Type specimen described by Constantine Samuel Rafinesque-Schmaltz in 1820,
collected in small brooks from the northern parts of New York State

Total length: metamorph = 1.0 inch (2.5 cm) to adult = 5.6 inches (14.1 cm)

The Northern Dusky Salamander is the most widely distributed *Desmognathus* in the Northeast, found in essentially every part of the region. It is the only Dusky Salamander that occurs in New England. Like other "duskies," it has large, stout hind legs that are built for jumping. When startled or captured by a human observer, it will make surprising leaps to escape, and if a predator or person is holding it by the tail, the body can escape while leaving the tail behind.

In most of its Northeast range, the Northern Dusky Salamander is likely to be confused only with the Allegheny Mountain Dusky Salamander or the Northern Two-lined Salamander. The adult Northern Two-lined Salamander has a yellow belly compared to the gray or brownish belly of the Northern or Allegheny Mountain Dusky. The Northern Dusky is more aquatic than the Allegheny Mountain Dusky and will rarely be found more than a short distance from the headwater stream and riparian zone. It has a keeled tail that becomes more pronounced in older individuals. Its back is decorated with random black spots, not lined with a row of chevron-like markings as on the Allegheny Mountain Dusky.

Because of its wide range and abundant habitat, Edward Drinker Cope suggested in the late 1800s that it was "perhaps the most abundant salamander in North America." In 1942, Sherman C. Bishop echoed that thought for New York, stating that the Northern Dusky Salamander "is perhaps the commonest and widely distributed species in the State." Bishop added, "In its somber markings and obscure pattern it is the least striking of all the species." In good habitat it is still locally abundant, but its status has undoubtedly changed over the decades. To thrive, it requires clean, flowing headwater streams and seeps surrounded by intact streamside habitat shaded by mature forests. Harvesting of trees for the forest industry and clearing of land for agriculture in the last century certainly modified many of these systems. Today, many of these areas are recovering as logging practices have improved and agricultural land is abandoned, but threats from rural developments and mining, especially mountaintop mining, continue. Unknown effects from stressors such as acid rain, atmospheric deposition of mercury, climate change, and introduced pathogens may result in robust populations of a widespread and common species to become a species of concern. This is a lesson learned recently, when White-nose Syndrome devastated what were once considered large, healthy populations of bats, changing their status from secure to endangered in just a few short years.

Allegheny Mountain Dusky Salamander
Desmognathus ochrophaeus

Type specimen described by Edward Drinker Cope in 1859, collected in
Susquehanna County, Pennsylvania

Total length: metamorph = 0.7 inches (1.7 cm) to adult = 4.4 inches (11.1 cm)

The Allegheny Mountain Dusky Salamander is one of nine species of Dusky Salamanders found in the Northeast. Identifying these species has been compared to working with what Roger Tory Peterson referred to in his *Field Guide to the Birds of Eastern and Central North America* as the "confusing fall warblers"—maybe even worse! And even seasoned observers may occasionally confuse a Northern Two-lined Salamander with an Allegheny Mountain Dusky Salamander. But the reward is that once you are able to confirm that you have found a population of Allegheny Mountain Dusky Salamanders, they are usually numerous, abundant even. They can be found any time of the year if you are willing to put in the effort and brave the chilly waters of the seeps and springs or the wet shale banks bordering the streams where the adults congregate to overwinter. There are still life history questions about this species that you can pursue.

The first step in determining the species is to check a field guide to determine which species might be found in your area. Six of the nine northeastern *Desmognathus* species are found only in Virginia or West Virginia. The Allegheny Mountain Dusky is not found in New England, only the Northern Dusky is. In the southern portion of the Northeast, the Allegheny Mountain Dusky is restricted to higher-elevation areas. To confuse things even more, where the range of Northern and Allegheny Moun-

tain Dusky overlap they sometimes hybridize, resulting in offspring that have characteristics of both species.

The Allegheny Mountain Dusky is a more terrestrial species than the Northern Dusky. The tail shape helps separate the two. The base of the tail is more round on the Allegheny Mountain Dusky, while the Northern Dusky, a more aquatic species, has a somewhat keeled tail that is better adapted for a swimming lifestyle. If you gently use your thumb and forefinger to hold both species about one-third of the way down their tail, you can roll the species with the round tail, but not the one with the keeled tail.

Surprisingly, reproduction by Allegheny Mountain Dusky Salamanders is not as well understood as one might expect. They appear to breed in both the spring and fall, and perhaps throughout the summer. The female guards the eggs for a period of seven to ten weeks until they hatch. Eggs deposited during ideal conditions may have larval periods of as little as one to three weeks. Eggs deposited late in the year, from the end of September to mid-October, or in stressful environmental conditions, will not transform for four to six months. In either case, they grow little during the larval period, adding just a few millimeters to their total length before transforming into the terrestrial form.

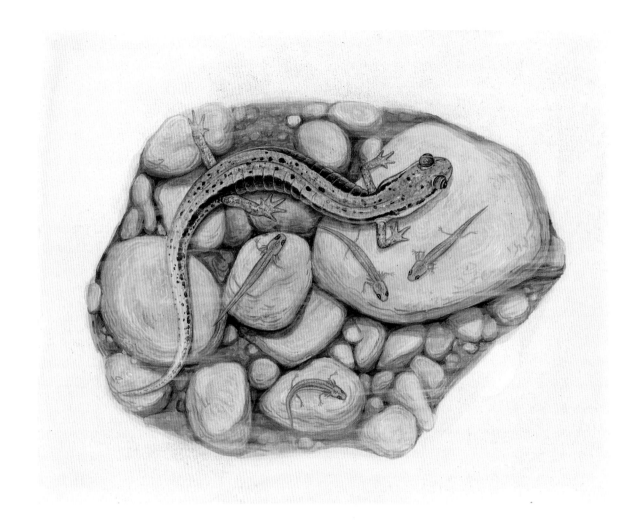

Northern Two-lined Salamander
Eurycea bislineata

Type specimen described by Jacob Green in 1818, collected from the vicinity of Princeton, New Jersey

Total length: metamorph = 1.7 inches (4.3 cm) to adult = 4.8 inches (12.1 cm)

The two lines of the Northern Two-lined Salamander are the two dark brown bands that border the light-colored broad band down the middle of the back. The vertically flattened tail is slightly longer than the body. The yellow belly of the adult Northern Two-lined Salamander separates it from the other *Eurycea* in our area and from the *Desmognathus* species, with which it may be confused. The Northern Two-lined Salamander is found throughout the region except for southern Virginia and southern West Virginia, where it is replaced by the Southern Two-lined Salamander (*Eurycea cirrigeria*), and a small area of the Blue Ridge in Virginia, where instead one finds the Blue Ridge Two-lined Salamander (*E. wilderae*). Separating these species may require counting costal grooves. The Northern Two-lined Salamander has 15 or 16 costal grooves, whereas the Southern Two-lined Salamander has only 14. The Blue Ridge Two-lined Salamander has 14 costal grooves if the elevation is below 4,000 feet (1,220 m), and 15 or 16 if above that elevation. It is best to check range maps in a field guide, because their areas of overlap are minor.

The Northern Two-lined Salamander rarely moves far from the streamside or seep areas. During daylight hours it is usually hidden under rocks, debris, or leaf litter along the edges of the waterway. It may be found in larger streams or rivers or in areas near where seeps or springs feed directly into lakes and ponds. This salamander nests under rocks in fast-moving, well-oxygenated water, where the female guards the eggs until they hatch after 30 to 70 days in late summer. The larvae take two years (in the north country, three years) before they transform into the adult form. The larvae can survive in ephemeral streams that dry up completely on the surface by retreating into the spaces between the rocks and gravel of the streambed where flowing water is always present. In cold-water lakes, Northern Two-lined Salamanders have been found nesting at depths of greater than 30 feet (9 m). In quiet pools, larvae can be observed at any time but are much easier to observe at night with a flashlight. As is typical of salamanders that breed in well-oxygenated streams, the external gills are quite small compared to those of pond-breeding salamanders such as the ambystomids.

The Northern Two-lined Salamander is a slender species with short, thin legs. It moves in a flattened, rapidly undulating snake-like manner when attempting to flee from predators, small children, or herpetologists. On rainy nights they can be found several hundred feet from the stream bank, which may result in them moving across road surfaces as they forage for small invertebrates. Because their movements from their typical habitat are small compared to the breeding migrations of spring-breeding amphibians, they are less likely to end up as roadkill, but because of their small size, they are less likely to be noticed.

Eastern Long-tailed Salamander
Eurycea longicauda longicauda

Type specimen described by Jacob Green in 1818, collected from the vicinity of
Princeton, New Jersey

Total length: metamorph = 1.7 inches (4.4 cm) to adult = 7.8 inches (19.7 cm)

Threatened in New Jersey

The Eastern Long-tailed Salamander is strikingly different from most other *Eurycea*, with its beautiful coloring and its surprisingly long tail. This yellow, sometimes yellow-orange or orange-red slender salamander has dark vertical herringbone markings on the sides of its long, vertically compressed tail. Dark blotches are scattered the length of its back. The tail may be nearly twice as long as the snout-to-vent length (SVL).

The SVL is a common method of measuring salamanders, frogs, lizards, and snakes. It doesn't apply to turtles, which can stretch out their necks or pull their heads into their rigid shells, significantly changing the distance from tip of snout to vent. The SVL is measured from the tip of the snout to the anterior end of the vent or cloaca opening. For frogs, the SVL is the same as the head–body length and is more properly called the snout–urostyle length (SUL). The SVL is slightly different from what is called the standard length (SL) in salamanders, where the vent is a slit that runs parallel to the length of the animal. For salamanders, the SL is measured from the tip of the snout to the posterior end of the vent opening. For smaller salamanders, the difference may only be a millimeter; in larger species, such as Hellbenders and Mudpuppies, the difference can be greater than a centimeter. In lizards and snakes, the vent is perpendicular to the body axis so that anterior and posterior ends of the vent are irrelevant, making the SVL equal to the SL. Care must be taken to record which measurement was actually taken so that meaningful comparisons can be made.

The Eastern Long-tailed Salamander may be confused with the Cave Salamander (*Eurycea lucifuga*) where their ranges overlap along the Ridge and Valley and Blue Ridge sections of Virginia and West Virginia. The Cave Salamander lacks vertical bars on the tail. Both species are found in limestone areas or at least in areas with a higher, more alkaline pH. Both species may be found in the twilight zone of cave openings, but in these situations there is normally only one species in a cave, either the Long-tailed or Cave Salamander, not both.

The Eastern Long-tailed Salamander is absent in New England, northern New York, and the coastal areas from Long Island to the Chesapeake Bay. The Long-tailed Salamander is a highly terrestrial species, spending most of its time throughout the year in upland hardwoods hiding under rocks, logs, or loose shale banks, often considerable distances from streams or spring seeps. The females move into the water to nest in late winter or early spring, depositing eggs attached to the undersides of rocks in seeps or small, flowing rivulets.

Spring Salamander
Gyrinophilus porphyriticus

Type specimen described by Jacob Green in 1827, collected from French Creek
near Meadville, Crawford County, Pennsylvania

Total length: metamorph = 3.9 inches (10.0 cm) to adult = 9.1 inches (23.2 cm)

Threatened in Connecticut

This light pink to salmon-colored salamander used to be called the Purple Salamander, a name that seems inappropriate for a living Spring Salamander. Its back and sides are covered with a pattern of darker reticulations, giving it a net-like appearance. As it ages, the mottling takes on an unwashed aspect, making the pink color perhaps vaguely purplish, but for me that is a stretch. A light stripe, most obvious in younger animals, runs from the eye to the nostril and then drops to the upper lip.

The Spring Salamander is the most aquatic of the lungless plethodontids. Its preferred habitat is cool, clear, spring-fed seeps and steep-gradient headwater streams surrounded by hardwood or mixed conifer-hardwood forests. They are also found in deep, shaded ravines and near the mouths of caves where subterranean streams discharge to the surface. Spring Salamanders are intolerant of warm, muddy waters or polluted streams. It is unlikely to find them far from the stream or wetland edge and almost never crossing a road, except during torrential rainstorms.

Larval Spring Salamanders emerge from their eggs at barely 1.0 inch (2.6 cm) in length, which is actually rather big for a newly hatched plethodontid salamander. Being a species of cool, well-oxygenated water, their external gills can be greatly reduced compared to those of pond-dwelling salamanders. Spring Salamanders are slow to develop, taking three years to transform into an adult form. As such, they can only successfully colonize and reproduce in permanent streams, but these streams can be subsurface streams. If the water flow is adequate, the dissolved oxygen concentration high enough, and the spaces between the rock rubble sufficient, Spring Salamanders can successfully colonize what appears to humans on the surface as an ephemeral stream.

Spring Salamanders feed on all sorts of aquatic invertebrates but will take small larval Dusky Salamanders, Two-lined Salamanders, Four-toed Salamanders, recently metamorphosed Wood Frogs, or even cannibalize larval Spring Salamanders. On rainy nights they can be seen lying quietly along the stream or wetland banks waiting in ambush for prey. When startled, they will take to the water and swim in an undulating eel-like fashion, with their legs pressed close to their bodies. They are less active but do continue to feed throughout the winter.

Where their ranges overlap, Spring, Red, Two-lined, and Northern Dusky Salamanders may be found occupying the same spring seep areas. These four species can coexist just a few feet apart, but if given the chance, the Spring and Red Salamanders will prey on the other two smaller species.

Mud Salamander
Pseudotriton montanus

Type specimen described by Spencer Fullerton Baird in 1849, collected from South Mountain, near Carlisle, Pennsylvania

Total length: metamorph = 2.6 inches (6.6 cm) to adult = 8.1 inches (20.7 cm)

Endangered in Pennsylvania

Threatened in New Jersey

The Mud Salamander in the Northeast exists as two subspecies separated by more than 90 miles (145 km), although their separation is only about 20 miles (32 km) farther south between the North Carolina and Tennessee populations. The Midland Mud Salamander (*Pseudotriton montanus diasticus*) is found in western Virginia and West Virginia. The Eastern Mud Salamander (*P. montanus montanus*) is found in eastern Virginia north to southern New Jersey. Their range overlaps in part with that of the Northern Red Salamander. The best feature to look for if unsure of the species is the color of the eye. The Mud Salamander has a mud-brown eye. The Northern Red Salamander has a yellow eye. The Mud Salamander is basically red with circular black dots, but usually not as many dots as on the Red Salamander. The red color of the eastern subspecies of the Mud Salamander is also a muted muddy-brownish wash on top of the red, whereas the Midland Mud Salamander is a brilliant red, much like a Northern Red Salamander.

The habitat of the Mud Salamander is not the clear, flowing water we associate with other salamanders that live in small streams and headwater spring seeps. Instead, the Mud Salamander is found in the muddy seeps formed from fine silt and decomposing organic matter along small streams, seeps, ditches, or floodplain ponds. These are habitats where the water flow can best be described as sluggish. Mud Salamanders burrow into this muddy ooze both to feed and to escape potential predators.

The females lay eggs in the autumn to midwinter, with the embryos hatching during the winter. Mud Salamander larvae are slow to mature, taking about one and a half to two and a half years to transform into the subadult. Perhaps because its habitat is not the faster-flowing oxygen-rich streams of other members of this guild, the Mud Salamander larvae have much more branched, bushy external gills than might be expected of a *Pseudotriton*.

When threatened, Mud Salamanders assume the defensive posture exhibited by a number of other species. They coil into a U-shape and tuck their noses beneath their bodies. They then raise their hind legs to lift their rear ends into the air, and curl their tails above their heads. This posture, along with the red body color, is thought to be a warning signal to potential predators that the intended prey may be poisonous or at least distasteful.

Queensnake
Regina septemvittata

Type specimen described by Thomas Say in 1825, collected from
Carlisle, Pennsylvania

Total length: at birth = 6.8 inches (17.3 cm) to adult = 36.3 inches (92.1 cm)

Endangered in New Jersey and New York

The Queensnake is a medium-sized, highly aquatic snake most closely related to watersnakes. In the early 1900s, the Queensnake and watersnakes were both included in the genus *Natrix*, a genus that is now restricted to Old World species. The New World species were placed in one of three genera: *Regina*, the Crayfish Snakes; *Nerodia*, the North American Watersnakes; and *Seminatrix*, the Black Swampsnakes.

A slender snake with keeled scales, the Queensnake has four brown stripes on its belly, with a yellow band about one and a half scales wide along its lower side separating the ventral side from a darker brown dorsum. A snake with four brown belly stripes can only be a Queensnake. Three thin dark lines run down the back, but they are hard to see except on an individual that has recently shed. The Queensnake's species name comes from the Latin *septem*, meaning "seven," and *vittata*, meaning "lengthwise stripes." It is a seven-striped snake: three on the back and four on the belly. The genus name *Regina* is Latin for "queen."

The Queensnake is not venomous or aggressive. I have never had one try to bite me; however, they may excrete a foul-smelling musk if handled. For some unexplained reason, the Linus character in the *Peanuts* comic has a fear and hatred for Queensnakes, believing them to be venomous and quite ugly. It is unfortunate that popular culture would paint such an inappropriate picture of such an interesting, inoffensive, and, I believe, quite attractive snake.

The Queensnake is rarely found far from water and even more rarely far from crayfish, its preferred food. Not only does it almost exclusively eat crayfish, but it also carefully selects crayfish that are soft shelled, the ones that have recently shed so that their new exoskeletons have not yet become hardened chitin. Eating a crayfish, even a soft-shelled one that is rather defenseless, is tricky. A crayfish's front pair of legs have powerful claws, so a cautious Queensnake will grab it from behind and swallow it whole.

This species is found in small creeks to medium-sized streams that are bordered by hardwoods. Queensnakes can be found under completely submerged rocks or those no more than a few feet from the stream's edge. They are diurnal and can be seen basking on rocks or on branches of shrubs overhanging the watercourse, even during the hottest days of midsummer. The most northern population of Queensnakes in the Northeast resides in a large wetland in Upstate New York that is also the home of a number of rare species, including Massasaugas, Coal Skinks, and, historically, Bog Turtles—an unusual species assemblage.

The only species that might be confused with Queensnakes are gartersnakes. The Short-headed Gartersnake looks most like a Queensnake from above but lacks the four brown stripes on the belly. Predators of the Queensnake include large fish and Black Racers. Perhaps ironically, Kingsnakes have been reported to eat Queensnakes.

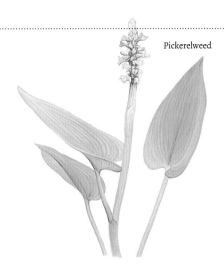
Pickerelweed

7 Small Waters

THIS HABITAT type is separated from seasonal wetlands by its permanent hydroperiod and from lakes by its depths and vegetation. Small waters include vegetated wetlands and ponds bordered by wetland vegetation. As a result, they support populations of fish, amphibians that spend more than one year as tadpoles or larvae, and basking turtles. These wetlands are traditionally called marshes, swamps, and beaver impoundments, and include both natural and man-made ponds. In the Northeast, 89 out of the 162 native species of herpetofauna use these wetland systems on a regular annual basis.

To correctly define these habitat types, a marsh is a wetland dominated by nonwoody vegetation, such as cattails (*Typha* spp.), sedges (*Carex* spp.), wetland grasses, rushes (*Juncus* spp.), and other herbaceous plants. A swamp is a wetland dominated by trees and taller shrubs such as Red Maple (*Acer rubrum*), elm (*Ulmus* spp.), alder (*Alnus* spp.), and other deciduous or coniferous trees and shrubs. Marshes and swamps may border or merge with shallow ponds, creating even more diverse habitats for amphibians and reptiles.

The shallow, open-water areas usually contain emergent vegetation that is rooted in the pond bottom and extends above the water surface, such as cattails, Pickerelweed (*Pontederia cordata*), bulrushes (*Scirpus* spp.), wild rice (*Zizania* spp.), and sedges. The slightly deeper zones, up to about 7 feet (2 m), are vegetated with submergent plants such as Coontails (*Ceratophyllum demersum*), waterweeds (*Elodea* spp.), milfoils (*Myrio-*

phyllum spp.), and pondweeds (*Potamogeton* spp.), or floating vegetation such as water lilies (*Nymphaea* spp.), bullhead lilies (*Nuphar* spp.), and free-floating species like duckweeds (*Lemna* spp.) and bladderworts (*Utricularia* spp.). The depth at which submergent species can be found varies with the turbidity of the water. In crystal-clear ponds or lakeshores, submergents may be found at depths that exceed 20 feet (6 m), but those conditions are exceptions rather than the rule. Both emergent and submergent aquatic vegetation provide protective cover for amphibian eggs, larvae, and adults.

Bladderworts deserve a special mention in these ecosystems. Bladderworts are carnivorous plants with an extensive underwater network of modified leaves. The only part that appears above the surface is the flowering stalk with a single or small cluster of yellow or purple flowers. The underwater leaves contain numerous small bladders that trap aquatic invertebrates. In one study, approximately 80% of bladderwort bladders contained mosquito larvae. A healthy population of bladderwort can reduce mosquito populations. Bladderworts combined with salamander larvae that also prey on mosquito larvae can help keep mosquitoes at a minimum without using chemicals.

Following centuries of unregulated harvest and tremendous declines, during the last century, Beaver (*Castor canadensis*) have made an extraordinary comeback throughout North America. Using their dam-building expertise, they have created numerous small ponds, shallow marshes, and open-canopied areas in the adjacent uplands. This activity has created prime breeding, nesting, basking, and overwintering habitat for many species of amphibians and reptiles. But their dam building has come into conflict with human use of the landscape. Many roads, homes, and agricultural fields have been built in low-lying areas during times when Beaver populations were at a minimum. When the Beavers return to these areas, however, flooding becomes more prevalent.

Red-spotted Newt
Notophthalmus viridescens

Type specimen described by Constantine Samuel Rafinesque-Schmaltz in 1820, collected from Lake George and Lake Champlain, New York

Total length: metamorph = 1.4 inches (3.5 cm) to adult = 5.5 inches (14.0 cm)

State amphibian of New Hampshire

The juvenile, or Red Eft stage, of this species is the most seen and well known of all salamanders in the Northeast. The Red Eft is entirely terrestrial. The skin of the Red Eft is less permeable than the skin of other northeastern salamanders, which better protects it from desiccation, and hence it can be found wandering on the surface almost any time during the warmer months. Its bright orange to red coloration with bright red spots circled in black is a warning to most potential predators that it is distasteful, even poisonous, if they try to eat it. The toxin, tarichatoxin, is poisonous and possibly deadly to some would-be predators. A few species have developed tolerance to the toxin and prey on Red Efts, including Common Gartersnakes, Hog-nosed Snakes, and Wild Turkeys, to name a few. But most mammals and fish find the Red Eft or the adult Red-spotted Newt distasteful. Several hundred Brook Trout (Salvelinus fontinalis) examined during a 10-year study found that not a single fish had ingested a Red-spotted Newt. Unlike many other salamanders, it is possible for Red-spotted Newts to successfully breed in ponds that contain predatory fish like Bass, Perch, and Pike. One may also find adult newts visiting fish-free ponds where Wood Frogs and ambystomid salamanders breed. At these ponds the newts feed on the egg masses of the other amphibians.

The Red Eft is slow to develop into an adult, taking up to seven years from the time the larva metamorphs into the land form until it returns to the pond to breed as an adult. The change from the eft stage to an adult is accompanied by more changes. Its bright orange back turns to a drab olive or yellowish-brown color but maintains the red spots. The belly changes to yellow with small black dots. During breeding season, the males develop a deeply keeled tail and dark, warty, black herringbone-like marks on the inner thighs. The females lack these dark markings on the thighs and have a much less markedly keeled tail. Once transforming into adulthood, Red-spotted Newts may remain in the pond for the rest of their lives. Newts remain active in the pond all winter, continuing to feed at a reduced level. As expected, with the colder water, their digestion and metabolism slow down a bit, so they eat less. When the pond is frozen, Newts can be seen swimming below the ice. Newts may be caught in minnow traps set by ice fishermen to catch bait.

The color variations and changes in body shape exhibited by the juvenile Red Eft and the adult male and female Red-spotted Newts during and after transforming back to an aquatic form have confused scientists in the past. James Ellsworth DeKay in his 1842 treatise on the reptiles and amphibians of New York treated the Red-spotted Newt as three separate species in two genera.

Eastern Cricket Frog
Acris crepitans

Type specimen described by James Ellsworth DeKay in 1842, collected from
The Locusts, near Oyster Bay, Nassau County, New York

Head–body length: metamorph = 0.6 inches (1.6 cm) to adult = 1.4 inches (3.5 cm)

Endangered in New York and Pennsylvania

The smallest of the frogs in the Northeast, an adult Eastern Cricket Frog weighs only 0.03 ounces (0.9 g). That would be about 500 frogs per pound! In contrast, the largest frog, the American Bullfrog, can weigh as much as 1.1 pounds (0.5 kg). The Eastern Cricket Frog is also the frog with the greatest color variation. Its skin is covered with small bumps the size of pinheads, giving it a warty appearance when examined closely. A dark triangular spot on the top of its head with the apex pointing toward the posterior helps identify this species. A band of color borders both sides of the triangle, starting on the nose and continuing down the back for the length of the body. This stripe can be a slightly darker brown than the sides of the frog but can also be tan, gray, or several shades of green, red, maroon, orange, or yellow. There is a dark stripe on the inner thigh that is hidden when the frog sits.

Acris crepitans is an Olympic-caliber jumper, capable of leaping more than 25 times its body length. With its light-weight and almost fully webbed hind feet, it can make a series of these jumps across the surface of the water without even breaking the surface tension on the water. Its breeding call, a series of crisp *gik-gik-giks*, is an in-sect-like clicking that sounds like two large pebbles being smacked together. Males call from the base of emergent vegetation, on floating vegetation, while floating in the water, or from the shore sporadically throughout the day, with the strongest choruses occurring after dark. Their diet is small insects, including mosquitoes, so it is a desirable species to have nearby. Cricket Frogs are in the treefrog family, but they lack the expanded toe pads of the members of the *Hyla* genus. As a result, they are a treefrog that does not climb trees. Their elevated perch in low vegetation can more often be obtained by jumping rather than climbing.

The Eastern Cricket Frog overwinters in deep underground soil fissures or rock crevices. In the fall, the adults and recent metamorphs can be found congregating near or on the shore before beginning their foray to their upland hibernacula. This migration may be a short jaunt or as much as 800 feet (244 m) from their breeding ponds, depending on the habitat surrounding the pond. Stand quietly in the woodlands in late September or early October and watch as Eastern Cricket Frogs are all hopping in the same general direction away from the pond. Repeat in the spring, and the movement is toward the pond.

American Bullfrog
Lithobates catesbeianus

Type specimen described by George Shaw in 1802, collected from the vicinity of Charleston, South Carolina

Head–body length: metamorph = 2.8 inches (7.1 cm) to adult = 8.0 inches (20.4 cm)

Generally just called the Bullfrog, the complete common name of this species is the American Bullfrog to distinguish it from large frogs in other countries that are also referred to as bullfrogs. Vernacular names can be so misleading and confusing. In Jamaica and the Philippines, the non-frog-like marine or cane toad (*Rhinella marina*) is called the bullfrog. The American Bullfrog is the largest frog in North America, with "bull" referring to its deep-bass advertisement call, *jug-o-rum*, that some observers have compared to the bellow of a bull, repeated slowly with pauses.

Bullfrogs produce tadpoles that overwinter one or two times, producing metamorphs in their second or third summer before leaving their breeding pond, lake, or slow-moving stream. The tadpoles are also the largest in North America, with a body size that barely fits in your hand and a total length of 6.7 inches (17.0 cm), which is considerably longer than the Bullfrog's metamorph, but then the tadpole's tail is resorbed during the transforming process.

Because of its size, the American Bullfrog is prized in the food industry for its legs, considered a delicacy in French and Asian cuisine, but it is also commonly eaten by hunters and fishermen who take them from the wild. The first day of frog season is not as celebrated as the first day of big-game hunting season, but those that enjoy frog legs look forward to that first day with a lot of enthusiasm. Although a number of species play a role in the international frog-leg market, the American Bullfrog is the key player in North America. Asian markets, most notably the Chinese, have been farm raising American Bullfrogs and then selling both live Bullfrogs and processed Bullfrog legs back to the Americans. An unofficial survey of Asian food markets in the capital district of New York found 25 markets selling live Bullfrogs imported from China, many of them suffering from the highly infectious Red Leg Disease. If these live frogs were purchased by someone who then returned them to the wild, those wild populations could also be infected.

Bullfrogs can be invasive. They quickly colonize newly constructed farm or ornamental ponds in residential areas, and they will colonize natural isolated fish-free ponds, especially when the forest canopy has been removed. Bullfrogs eat anything they can get in their mouths, meaning other frogs, salamanders, turtles, birds, bats, small mammals, and even other Bullfrogs. Accidently or intentionally released American Bullfrogs quickly became a problem in the western United States and Canada, where they added one more stress to already imperiled species of amphibians. They have become serious pests in a number of foreign countries where they have been released. Even in areas where they are a native species, Bullfrogs can be a problem. Analysis of the stomach contents of Bullfrogs in Massachusetts demonstrated that they are one of the main predators of hatchling Plymouth Red-bellied Turtles, a federally endangered species.

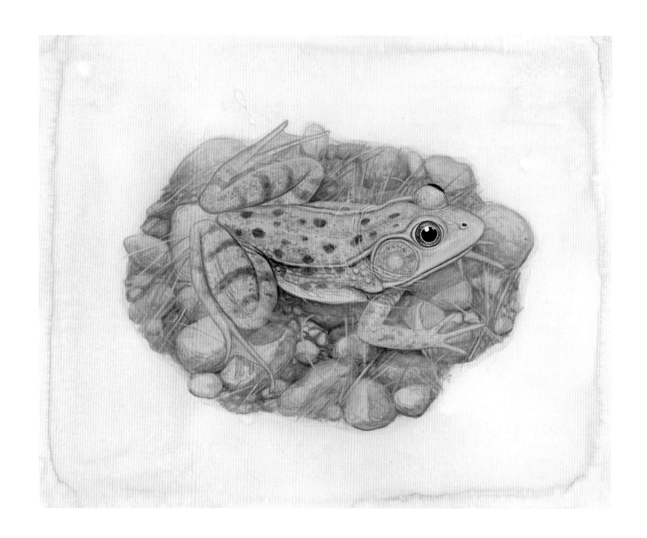

Green Frog
Lithobates clamitans melanota

Type specimen described by Pierre André Latreille in 1801, collected from
Charleston, South Carolina

Head–body length: metamorph = 1.2 inches (3.0 cm) to adult = 4.3 inches (10.8 cm)

Gunk. Gunk. Gunk. The Green Frog's distinctive call, like a plucked out-of-tune banjo, is a welcome sign of summer. Found in all habitats except the brackish coastal areas and the interiors of dry pine-oak forests, this frog's life centers on the ponds, lakes, wetlands, and rivers of the Northeast. The calling stops or at least diminishes as you approach a pond when the Green Frog chorus is in full swing. Come a little closer, and you will hear the warning call followed by a splash as the startled frog leaps from its perch on shore to the safety of the water.

With a size that overlaps that of a younger Bullfrog and a matching randomly mottled or spotted green to greenish-brown back and a white belly, a closer look reveals the field characters that separate the two. The Green Frog has a distinct dorsolateral ridge running down each side of its back, starting at the eye and extending nearly to the hips. The dark spots on the hind legs, less distinct on the Bullfrog, form bars that when the legs are folded appear as crossbands that line up on thigh and lower leg. Whereas the Bullfrog's underside may have a pale yellow wash, on the adult male Green Frog the chin exhibits a bright yellow color, a trait that is easy to see on a frog floating many feet off shore.

The yellow throat is a hint to the green of the Green Frog. Green Frogs actually have no green pigments in their skin. The color is a result of a mix of the specialized cells in the skin called chromatophores. In the case of frogs that appear green, the uppermost layer of chromatophores, the xanthophores, contain yellow pigment. Under the xanthophore layer are the iridophores, which reflect light, giving the skin a blue appearance. When the two are combined, the appearance is green. When the yellow is missing, the frog's skin appears to be blue, and when the blue is missing, the skin appears yellow. The chin of the male lacks the blue, and the chin of the female lacks both the blue and the yellow. A third group of chromatophores, the melanophores, are responsible for dark brown or black markings on the frog.

Green Frogs breed in early summer, laying egg masses that form a thin sheet of eggs on the water's surface. Within a day or two, this sheet breaks up and settles on the pond bottom. Tadpoles take two years to transform and leave the pond, so they won't be found breeding in the seasonal pools that dry each year. Green Frogs are content to spend their summer around breeding ponds, but during periods of wet weather they explore surrounding uplands searching for new ponds to colonize. Build a new pond, and if Green Frogs are within a couple thousand feet, they will find it. In early fall, Green Frogs leave the quiet ponds to overwinter in the feeder streams that enter the breeding pond. In midwinter, the frogs are in retreat, hiding under rocks or logs in shallow streams that don't freeze but because of the flowing water are better oxygenated than the muddy pond bottoms where they spent the summer.

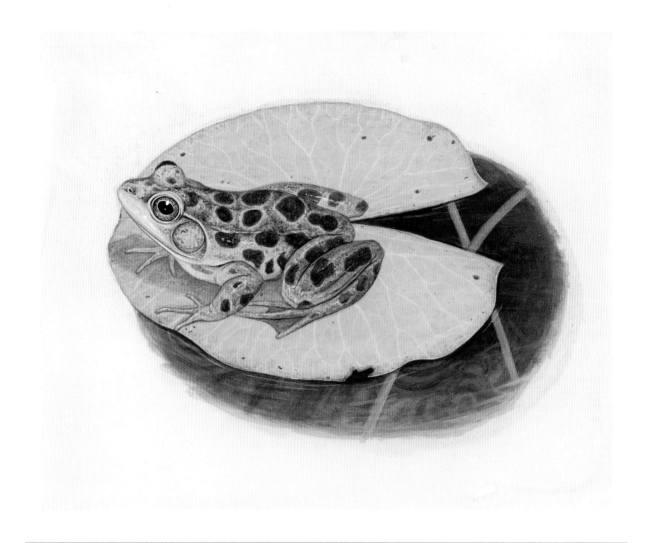

Mink Frog
Lithobates septentrionalis

Type specimen described by Spencer Fullerton Baird in 1854, collected from
Lake Itasca, Minnesota, and Sacket's Harbor, New York

Head–body length: metamorph = 0.4 inches (1.1 cm) to adult = 3.0 inches (7.6 cm)

I was leading a field trip to a wetland bordering a small lake in the western Adirondacks when we encountered several Green Frogs and the prized Mink Frog. The pattern of mottling on the dorsal surface of the Mink Frog is usually distinct but sometimes is similar to that of a Green Frog. And to confuse issues even more, the pattern on a Green Frog in these northern climes sometimes makes it look like a Mink Frog. So it was a perfect chance to do a side-by-side comparison.

The first characteristic is the pattern on the hind legs. On the Green Frog, there are distinct bands that run perpendicular to the leg. When folded, as if the frog were getting ready to jump, the bands on the upper leg line up pretty well with the bands on the lower leg. The pattern on the legs of the Mink Frog is more random, not creating a coherent pattern at all. The group was beginning to see the difference between these two species.

The second characteristic is the smell, which is how the Mink Frog got its name. The smell of a Mink Frog has been compared to that of a Mink, a member of the weasel family, Mustelidae. The family name says it all. A musty odor is common to all weasels, some more than others. But it is a poor comparison to use on this frog, for how many of us have actually smelled a Mink? Yet anyone who has spent time afield is familiar with its more odoriferous cousin, the Skunk. A calm, resting Mink Frog produces little or no discernible odor, so I wanted to annoy this one just enough to identify it from a Green Frog by the nose alone—our noses, not his. I gently picked it out of the container and held it by gripping around its waist with a moist hand, keeping its legs hanging down. Then I allowed each member of the group to get close enough to smell it. Most of them immediately withdrew with a wrinkled nose, indicating they had received the message. But one-third of the class indicated that they could not tell the Mink Frog from the Green Frog by odor. That was a response I didn't expect, and I joked that a third of my class "did not smell good." But the lesson learned is that not everyone can detect that particular odor.

The third characteristic is the webbing on the hind feet, and it is a solid method of identifying a Mink Frog,

for there is no overlap here with the Green Frog. Holding a frog and getting it to cooperate by spreading its toes so you can determine how much webbing is between the toes on the hind feet causes both you and the frog a bit of distress. I don't suggest trying it. Instead, place the frog in a container of water deep enough so that it has to float on the surface and yet can't climb out. Unlike treefrogs, Mink Frogs and Green Frogs cannot climb the smooth surface of a vertical-sided bucket and will soon relax and just lay motionless. Then take a careful look at the webbing between the toes of its hind feet. On the Mink Frog, the webbing extends to the tip of the fifth toe, the one farthest from the body, and to the last joint of the fourth toe. The webbing on the fifth toe of the Green Frog does not extend to the tip of the fifth toe and barely extends beyond the second joint of the fourth toe.

The fourth and final characteristic is the voice. We were not at this wetland at the right time of year—June to July—to hear its call, but if you happen to be there during breeding season, the call is distinctive. No other northern frog has a call that sounds like a guttural *cut-cut-cut*.

The Mink Frog is the frog of the North. Although its range does not extend as far north as that of the Wood Frog or Northern Leopard Frog, its southern limit is farther north than that of other frogs. That may not bode well for its future as a frog of the northeastern United States. I associate Mink Frogs with small Beaver meadows and other impoundments where emergent and floating vegetation is prominent. Beaver impoundments and Mink Frogs disappear as we climb the higher mountains of northern New York and New England. Mink Frogs can't survive farther south, where the average July temperatures are above 66°F (19°C). If climate change predictions are accurate, this frog will be forced out of the southern reaches of its current range, but it won't be able to escape to higher elevations because forest succession at the higher altitudes won't be able to keep pace with the pace of its displacement. By the end of the twenty-first century, the Mink Frog may be gone from New York, Vermont, and New Hampshire, surviving only in northern Maine, the upper Midwest, and Canada.

Snapping Turtle
Chelydra serpentina

Type specimen described by Carolus Linnaeus in 1758, collected from the vicinity of New York City

Carapace length: hatchling = 0.7 inches (1.9 cm) to adult = 19.4 inches (49.4 cm)

State reptile of New York

Prehistoric in appearance with a defensive attitude, the Snapping Turtle is the largest freshwater turtle by weight and one that can give a rather nasty bite if approached from the wrong end. Actually either end, if you are not careful, because a Snapping Turtle can turn quite quickly while perched, standing high on its hind legs. With its long neck and lightning-quick strike, it can catch an unwary potential predator with a nasty warning bite that says "don't mess with me!"

The Snapping Turtle is easily identified by its relatively flat, dark brown to nearly black carapace, which has three prominent ridges running its length. These ridges become less obvious with age. It has a disproportionately large head and a long tail, the length of the carapace or even longer, accented with large serrations or saw teeth along its upper edge. Flipped over on its back, the Snapping Turtle's greatly reduced plastron indicates that its shell offers less built-in protection from a predator than is found in most other turtles. This might help explain why it is always willing to put up a fight when threatened. It cannot simply retreat into its shell but has to defend itself with its strong bite and long claws.

Snapping Turtles are found in almost any aquatic system, from the slightly brackish tidewaters of estuaries to major rivers and lakes to small farm ponds and gently flowing Beaver ponds. Snappers are not compulsive baskers, as are our other lake, pond, and riverine turtles, but they may be seen basking on logs for short periods early in the spring. Their diet is truly cosmopolitan. Although despised in the past as a threat to game fish and waterfowl, in truth they are mostly satisfied with easy pickings such as carrion and lots of aquatic vegetation. They will take live fish, frogs, reptiles, mammals, and birds when the opportunity presents itself. Being an efficient underwater predator, it is surprising how Snapping Turtles react to humans who are swimming or wading in the water. Rather than striking or biting at human legs or toes, they prefer to quickly retreat into the mud or vegetation. That is not to say they won't bite if provoked, but it would be the exception rather than the rule.

If Snapping Turtles are reluctant to bite humans, at least when they are in the water, humans are not reluctant to bite Snapping Turtles. Snapping Turtle soup is a favorite among some groups, with the soup being especially popular in the Philadelphia area. For those who do eat Snapping Turtles, one caution: of all vertebrates tested to date, Snapping Turtles bioaccumulate higher levels of organochlorides—including DDT, polychlorinated biphenyls (PCBs), and other pesticides, some of which are known to cause reproductive problems in birds—than any other species. This body burden can be passed from the female directly to her offspring through the egg. In the past, tens of thousands of Snapping Turtles and their eggs have been harvested from every state in the Northeast, mostly for export to Asian markets. Most northeastern states allow personal and commercial harvest of Snapping Turtles.

Painted Turtle
Chrysemys picta

Type specimen described by Johann Gottlob Schneider in 1783, collected from the vicinity of New York City

Carapace length: hatchling = 0.9 inches (2.2 cm) to adult = 7.1 inches (18.1 cm)

State reptile of Vermont

The Painted Turtle is the most commonly seen basking turtle, especially in northern portions of its range. Many turtle species bask primarily in the spring, when the water temperature is in the low 50°F (10°C) range and the air is warmer or the sun is out. Once the water warms into the high 50s, there is much less basking activity, but the Painted Turtle will continue to bask throughout the summer and into fall, except when the air is cool and the sky is overcast.

Two closely related subspecies of the Painted Turtle are found in the Northeast: the Eastern Painted Turtle (*Chrysemys picta picta*) and the Midland Painted Turtle (*C. picta marginata*). As the names imply, the eastern subspecies occupies the eastern portion of our region, while the midland subspecies is found in the western sections, primarily areas within the Mississippi and Great Lakes drainages. There is a wide zone of overlap, and within that zone a lot of hybridization of the two subspecies occurs. The Eastern Painted Turtle has an unmarked yellow plastron, whereas the Midland Painted Turtle has an elongated and irregularly shaped dark blotch in the center of its plastron. The smooth, black- to olive-colored carapace on both subspecies is highlighted with bright red markings on the smaller, marginal scutes. The pattern of the larger scutes helps distinguish the subspecies. On the eastern subspecies, the posterior edge of the larger scutes on the carapace is basically a straight line with broad light-colored bands, the sutures, between all the large scutes. The midland subspecies has the normal staggered arrangement of the large scutes without the broad light-colored sutures separating them. Although this is an obvious character than can be seen at a distance through a spotting scope or binoculars, it does not always offer definitive identification. In the zone of hybridization, I have seen numerous Painted Turtles with a carapace that looks like one subspecies and a plastron that looks like the other.

This is complicated by the fact that an occasional individual of the southern or western subspecies may be found in the wild, and they too may hybridize with the native subspecies. Painted Turtles have been a popular species in the pet trade, and these nonnative subspecies may be the result of intentionally or accidently released pets. More recently, Painted Turtles have been showing up as a common species to stock in ornamental ponds. Many of these stocked turtles end up deciding that ornamental ponds are not the best habitat, and they wander off to settle in nearby natural ponds, if they are not killed by terrestrial predators or on roadways getting there.

For many turtle species, it is possible to determine what sex they are by examining their plastron. But that doesn't work for Painted Turtles because both sexes have nearly flat plastrons. Instead, male Painted Turtles have distinctly longer claws on their front legs than females. The claws on the females have a maximum length of about 0.4 inches (0.9 cm), whereas males' claws are much longer. The male uses these long claws to stroke the neck and head of the female while facing her as part of their courtship ritual. A second method of determining sex is to check the tail. As in most species of turtles, the male's tail is distinctly longer and fatter than the females, with the cloacal opening positioned beyond the edge of the carapace.

The date when Painted Turtles nest is directly related to latitude: early May in the South to mid-July in the North. The largest females can lay as many as 23 eggs. Hatchings emerge from the nest in 62 to 75 days, but many times they overwinter in the nest after hatching to emerge in the spring when the ground thaws. They are able to survive in the nests just a few inches underground because of an "antifreeze" in their tissues that prevents freezing of the cells followed by death.

Blanding's Turtle
Emydoidea blandingi

Type specimen described by John Edwards Holbrook in 1838, collected from the
Fox River, Illinois

Carapace length: hatchling = 1.0 inch (2.5 cm) to adult = 11.2 inches (28.4 cm)

Endangered in Maine and New Hampshire

Threatened in Massachusetts and New York

The prominent deep yellow chin sticking slightly out of the water about 50 feet (15 m) offshore caught our eye immediately. Perhaps it's a Green Frog. No, it's a Blanding's Turtle. The hallmark bright yellow chin and neck are the best field characters for identifying the Blanding's Turtle at a distance. When basking on a log, the streak of yellow sets it off from any other turtle. In profile, this turtle looks like an army helmet, a high-domed and elongated shell. Yellow spots, more pale than the yellow of the throat, are profusely scattered on the carapace.

Because it is has such a high-domed shell and is quite terrestrial, making long nesting migrations but also just moving from one small wetland pool to another, this turtle has been referred to as the Blanding's Tortoise. It has also been called the Semi-box Turtle because the flexible lobes on its plastron can be pulled in when the turtle retreats into its shell, but not as tightly as in a true Box Turtle.

In many populations, the female Blanding's Turtles travel 0.5 to 1.0 miles (0.8 to 1.6 km) from their home wetland to nest, generally seeking out the same nesting area each year. This means, especially in the Northeast, that the female must cross developed areas with roads. In more agricultural areas, she will have to cross hayfields about the time of first haying, in late May to mid-June, which coincides with their nesting period. Moving at a turtle's pace gets new meaning when you measure how long it takes for the females to make these journeys. A female may take two days to a week to move from the overwintering pond to the general nesting area, and then another day or two to decide exactly when and where to nest. After depositing her eggs, she then has to make the return trip crossing through the same hazardous areas.

The Blanding's Turtle's distribution is puzzling. The largest concentration of populations in the Northeast occurs along the St. Lawrence River in New York and is continuous with those in Canada that extend westward through the upper Midwest to the turtle's westernmost populations in Nebraska. Nebraska holds the mother lode. There is evidence that there are more Blanding's Turtles in Nebraska than anywhere else in its range, and possibly in the rest of the turtles' range combined. In the rest of the Northeast, there are scattered populations along Lake Erie, eastern and southeastern New York, the greater Boston area extending into southern New Hampshire and Maine, and then several extremely disjunct populations in Nova Scotia. Blanding's Turtles may have been extirpated from Pennsylvania, where they had previously occupied habitats in the northwest corner of the state.

Yellow-bellied Slider
Trachemys scripta scripta

Type specimen described by Johann David Schoepff in 1792, collected from the vicinity of Charleston, South Carolina

Carapace length: hatchling = 1.1 inches (2.8 cm) to adult = 11.4 inches (28.9 cm)

Pond Sliders are widespread and common turtles that were first described for science as *Testudo scripta* from a specimen collected in 1792. But where was it found? No one knows for sure, because that information was not recorded. Where is the specimen now, so today's scientists can examine the type specimen? In this case, that information is also not known. Since being first discovered, other specimens of the Pond Slider have been "first" collected, described, named, and renamed more than 40 times between 1792 and 1916. For scientists who deal with describing and naming organisms, it is the earliest published description that counts. Some of the extra names and confusion come from the variation exhibited by the turtle, resulting in the naming of three currently accepted subspecies. *Trachemys scripta scripta* (Yellow-bellied Slider) was the first described. *T. scripta troostii* (Cumberland Slider) and *T. scripta elegans* (Red-eared Slider) followed in 1836 and 1838, respectively. A herpetologist, Karl Schmidt, reviewing the data in 1953, made his best guess and designated Charleston, South Carolina, as the type location for *T. scripta*. This date and location also apply to the subspecies *T. scripta scripta*, but the other two subspecies have their own designated type specimens and type locations.

The importance of all this becomes apparent when we consider how scientists, conservationists, and wildlife managers use the information. Descriptions of life histories, behavioral traits, evolutionary relationships, genetics, physiology, conservation, and management options are all dependent on knowing that we are all talking about the same species or subspecies. If we are not talking about the same organism, then perhaps the conclusions are not valid. The Pond Slider represents an extreme example of being "first" described many times, but virtually all species suffer similar confusion.

All three slider subspecies occur in the Northeast. The natural range of the Red-eared Slider extends into western West Virginia. The Yellow-bellied Slider is found in southeast Virginia, and the Cumberland Slider occurs in southwest Virginia. The Red-eared Slider became the most popular pet turtle following World War II. Hatchlings were sold by the tens of thousands and introduced many children to the beauty of turtles. Probably most of these turtles died in captivity, because few people took the time to learn proper captive care of young turtles. Many were released to the wild, and some survived. Red-eared Sliders are now one of the most widespread species in the United States, showing up in many areas outside of their natural range. Not just showing up, but successfully overwintering in harsher climates north of their natural range and reproducing. They are also being exported to other countries in large numbers. Between 1989 and 1997, the United States exported over 52 million Red-eared Sliders to other countries. Red-eared Sliders have become established as wild populations in many areas in the United States, including most of the northeastern states as far north as Maine, and other countries where they outcompete native turtles for basking and nesting sites.

Common Rainbow Snake
Farancia erytrogramma erytrogramma

Type specimen described by Ambroise Marie François Joseph Palisot de Beauvois in 1801, collected from the lower Cooper River in the vicinity of Charleston, South Carolina

Total length: hatchling = 7.8 inches (19.7 cm) to adult = 66.0 inches (167.6 cm)

Endangered in Maryland

The multicolored, smooth-scaled Common Rainbow Snake does justice to its common name. One of the most colorful snakes in the Northeast, it is found in wetlands, rivers, streams, and lakes within the coastal plain and brackish marshes of eastern Virginia and southern Maryland. It is iridescent black with three prominent red stripes running down its back, and a generally red to pinkish belly with two rows of black dots. The top of its head is black, marked with bilaterally symmetrical red dots and squiggles. The lower sides of the snake below the red lateral stripe vary from dark violet to slate blue and tinged with yellow. At the end of its tail is a modified scale that forms a spine-like tip.

The Rainbow Snake's small head, barely wider than its body, is well suited for burrowing in the sandy uplands where it nests and overwinters. The female is quite a bit larger than the male and often remains with the nest while the eggs are incubating. During the first couple years of life, young Rainbow Snakes spend considerable time in the uplands and small wetlands feeding on frogs and salamanders, with tadpoles being the top menu item.

Adults of this highly aquatic species have adopted a nocturnal lifestyle, with a strong preference for feeding on American Eels (*Anguilla rostrata*), which are also highly nocturnal. Rainbow Snakes are active on the surface in uplands when it rains and have been observed taking eels out of water.

With eels being their favorite food, the health of the Rainbow Snake populations hinges on the health of the American Eel populations. Over the last century, American Eels have been declining owing to a variety of factors, including construction of dams that blocked their upstream migrations, overharvesting by humans for fish bait or as a valuable food source, getting sucked into the water intakes for power plants, or being run through turbines at hydroelectric facilities. Several populations of eels have showed signs of rebounding after the removal of a large dam in Virginia, which is positive news for the snakes. Because eels are such an important economic commodity to the human population, with young eels bringing several hundred dollars a pound, there are significant efforts to ensure that eels continue to thrive and thus so will the Rainbow Snake.

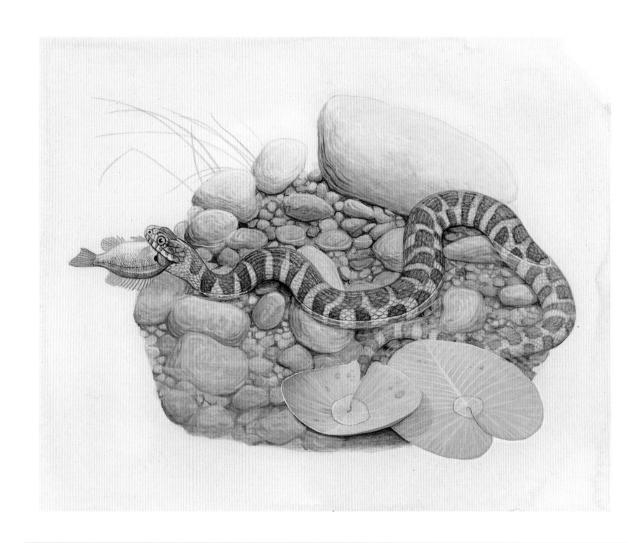

Northern Watersnake
Nerodia sipedon

Type specimen described by Carolus Linnaeus in 1758, collected from the vicinity of New York City

Total length: at birth = 7.5 inches (19.0 cm) to adult = 55.3 inches (140.5 cm)

If you live near or vacation on water in the Northeast, except for northern New York and the northern portions of New England, you will most likely encounter a Northern Watersnake. The Northern Watersnake can be found in almost any type of water body in the region, from slightly brackish coastal waters to all types of wetlands, as well as the flowing waters of rivers and streams to the quiet waters of ponds and lakes. Being such an aquatic species, it is only natural that its main foods are any fish or amphibian that it can capture and swallow whole. If the opportunity arises, it will also take some invertebrates and even small mammals such as voles and shrews.

Recreational boaters, swimmers, and fishermen who encounter a Northern Watersnake sometimes mistakenly report finding the venomous Eastern Cottonmouth or Water Moccasin (*Agkistrodon piscivorus*), which in truth only occurs as far north as southeastern Virginia. The aquatic habitat of the Northern Watersnake and its feisty disposition when startled or threatened combine to give this nonvenomous snake a bad reputation.

A careful examination of a few characteristics will help decide which of the two species you have encountered in the small area where their ranges overlap. You should be assured that you have nothing to fear in the rest of the region where only the Northern Watersnake is found. Both species are marked with crossbands. In the Northern Watersnake, the light-colored markings look like a series of squarish blotches running the length of its back and tail. The blotches at the head end of the snake extend onto the sides as a continuous saddle-shaped blotch. The spots toward the tail are staggered with side blotches that usually meet at the corners of the squares along the back, giving the snake a slight checkerboard appearance. On young snakes, especially those that have recently shed, the blotches are bright reddish to reddish brown with dark edges on a light gray background color. As the snakes mature or if they have not shed in a while, the pattern becomes obscure unless examined closely. The pattern shows up more readily when the snake is in the water than when it is observed basking onshore or atop a log. There are a few populations where the markings have all but disappeared completely.

The markings on the Cottonmouth are more distinct crossbands, but they may also become obscure with age and shedding condition. If threatened on land, the Cottonmouth retreats into a tight coil and holds its head almost vertically with its mouth spread wide, revealing the white "cotton" interior of the mouth that gives it its name. Whereas the bite of a Northern Watersnake can hurt and draw blood, it is not life threatening. The bite of a Cottonmouth can be serious, so in that small area of southeast Virginia where both species reside, be cautious when approaching a snake near the water that appears to be banded.

Crayfish

8 Big Waters

RIVERS AND lakes are the "big waters" of the Northeast and, next to the extensive northern woodlands, are the most well-known and enjoyed natural feature of the region. Big waters include streams large enough for recreational boating and fishing as well as the really big waters like the Chesapeake Bay and the Great Lakes. Both natural and man-made dams and reservoirs are included as well. Because of the aquatic nature of many of the herpetofaunal species in the Northeast, some species are restricted to just one or a few watersheds. There are three major drainage basins in the Northeast: the Ohio-Allegheny-Monongahela River drainage reaches the Gulf of Mexico via the Mississippi River; the Great Lakes–St. Lawrence River drainage reaches the North Atlantic in the Canadian Maritimes; and the mid-Atlantic area that includes rivers from the Penobscot in Maine south to the Susquehanna and Potomac Rivers flowing into the Chesapeake Bay, plus other major rivers such as the Hudson, Delaware, and Connecticut. Higher-gradient streams such as the upper Delaware, Susquehanna, and Penobscot Rivers are prime recreational areas for canoeists and kayakers.

Large and small lakes—some very deep and some quite shallow—interrupt the vast network of northeastern rivers. The deepest lake is Lake Ontario at 801 feet (244 m) deep with a surface area of 7,320 square miles (18,960 km²). Lake Erie is larger, but only a small portion of that Great Lake is within the Northeast region. Other deep lakes include the Finger Lakes in New York, which were formed following the last ice age. The deepest of these is Seneca Lake at 618 feet (188 m) deep. Note that Lake Ontario and Seneca Lake, as well as a few other deep lakes in the Northeast, have depths that place their bottoms below sea level. The largest lake in the Northeast, other than the Great Lakes, is Lake Champlain, which separates New York from Vermont with a surface area of 490 square miles (1,270 km²) and a depth of 400 feet (122 m).

Natural lakes are most abundant in the portion of the Northeast that was covered by glaciers during the most recent advance of ice sheets: the Wisconsinan Glacial Event, which reached its maximum about 22,000 years ago in Pennsylvania and New York. Areas to the south of the glacial maximum have fewer lakes, with West Virginia having no natural lakes at all. The abundance of waterways in the Northeast has attracted human population since long before European settlement. Waterways provided both a means of transportation and a food source, not to mention a potable water supply. In more modern times, the attractiveness of these waters for humans has been to the detriment of many native aquatic species. Threats include recreational canoeing and kayaking, commercial boat traffic, shoreline development, bulk heading of the shoreline, siltation, toxic chemicals, dredging, channelization, and loss of floodplain forests. In addition, the stocking of nonnative fish, the dumping of live bait that may outcompete native species, and the spreading of disease through the transfer of live fish and bait have had negative effects on some amphibian and reptile populations.

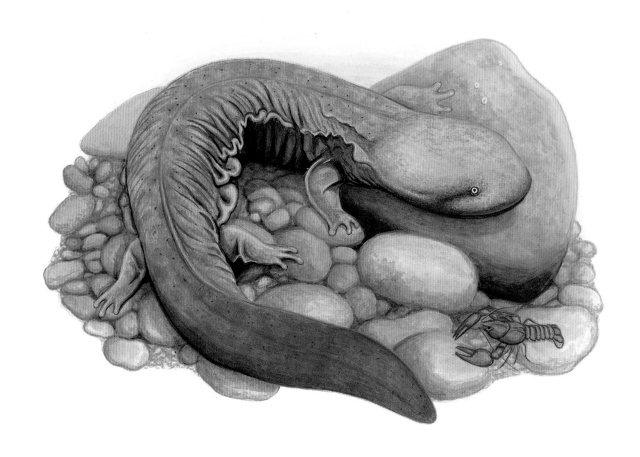

Eastern Hellbender
Cryptobranchus alleganiensis alleganiensis

Type specimen described by François Marie Daudin in 1803, collected from the Allegheny Mountains, Virginia

Total length: metamorph = 3.9 inches (10.0 cm) to adult = 29.0 inches (73.7 cm)

Endangered in Maryland

"Ugly." I have often heard that term used to describe the Eastern Hellbender. But I would say their appearance is more primitive than ugly. Hellbenders look unlike any species you have seen before: mud brown in color, loose folds of skin running down their sides, beady little eyes, and mouths as wide as their very wide heads. And if you see a full-grown adult, they are huge for an amphibian and possibly even older than you are! Unlike James Ellsworth DeKay, author of *The Zoology of New York*, who lamented in 1842, "I have never met with this animal myself," I was fortunate to meet the Hellbender courtesy of the late Professor Richard "Dick" Bothner of St. Bonaventure University during the summer of 1985. Careful rock lifting was our modus operandi. Dick was quick to scold if you lifted what he called a "sexy-look-ing" cover rock, incorrectly resulting in a dark cloud of suspended sediments that would hide the escaping Hellbender. And his scolding was even harsher if you did not carefully reposition the rock so it would remain suitable as a Hellbender shelter rock. To me, those first few Hellbenders were huge, larger than any salamander I had found before. But when we found the really big ones I was speechless. They were nearly 2 feet (61 cm) long and as big around as my forearm! Hellbenders are strong, muscular, slithery creatures that are hard to hold. A dip net and a large, flat-bottomed container were essential to get a good look and to give me an opportunity to wipe off the gobs of white slime the animal had exuded on me. As with most salamander slime, it really does not wipe off, nor does it stay white. Dark-stained hands often result from handling a Hellbender.

Hellbenders have declined to such an extent that there are few places left in the Northeast where they are easy to find. They are also one of the species that contradicts what it means to be an amphibian, living a life that is half aquatic and half terrestrial. Hellbenders are purely aquatic from egg to larva to adult, preferring cool, clear, fast-flowing, and well-oxygenated water—the kind of streams where I often find Wood Turtles. I think of the best habitat for Hellbenders to be trout-quality streams where their favorite food—crayfish—is abundant.

But many of these streams have disappeared because of a variety of land-clearing activities, dam construction, stream realignment, and road construction. Fishermen who fear the bite of the Hellbender or mistakenly think they are predatory on game fish will often kill them when they are accidently hooked. Today in the Northeast, Hellbenders are more frequently found in slower-flowing streams burdened with large amounts of suspended sol-ids. At these sites, reproduction is often poor, and what we are left with are often aged populations where most members are 25 or more years old.

Breeding season begins in late summer. At this time of the year it is easiest to find them as they congregate in areas where there are numerous rocks suitable for nesting under. It is also the time when the most harm can be done. If a nest rock is lifted, many of the eggs are washed loose, and it is impossible to get them back under the rock, where the male guards the eggs. And if the rock is not replaced properly, it may no longer be suitable as a cover or nest rock at all.

Since that first day with Dick, I have spent countless hours searching for them in the Susquehanna and Mis-sissippi drainages, more times than not ending the day without finding a single Hellbender. But one day stands out. Although their movements away from cover rocks are usually nocturnal, one fall afternoon we found what looked like a large male lying in plain view about 6 feet (1.8 m) from the closest cover rock. On closer examina-tion we saw it was not one but four individuals piled on top of each other. Within 10 feet (3 m) there were two more Hellbenders moving toward this pile. We did not want to disturb this gathering to determine size and sex of all the participants, so we just watched in fascination from a respectable distance. They were so absorbed in their own affairs that they did not notice us. At the end of one particularly successful day in the field, Dick com-mented about what he found so great about studying Hellbenders. "On the average, Hellbenders are uglier than we are," he said as he turned and glanced directly at me before adding "on the average." *Primitive-looking, intrigu-ing critters*, I thought.

Common Mudpuppy
Necturus maculosus

Type specimen described by Constantine Samuel Rafinesque-Schmaltz in 1818, collected from the Ohio River

Total length: larva = 0.9 inches (2.2 cm) to adult = 19.1 inches (48.6 cm)

Extirpated from Maryland

Introduced in Connecticut, Maine, Massachusetts, and New Hampshire

The Common Mudpuppy is a salamander that never grew up. At least not in the sense that we expect amphibians to develop: from egg to aquatic larva with external gills to terrestrial adult living on land. The Hellbender is similar, as it never leaves the water to live on land, but the Mudpuppy is unique among northeastern herpetofauna in that it keeps its external gills throughout its life. It is a large salamander and deservedly is a species of the big waters, generally larger rivers and lakes, although occasionally it is found in smaller streams and in the floodplain wetlands bordering rivers. Being completely aquatic, it has a hard time spreading to new drainage basins. In the Northeast, it is native to the Mississippi–Ohio River and the St. Lawrence–Great Lakes drainages. It is believed to have been accidently introduced into the Hudson River through the Erie or Champlain Canals and deliberately into the Connecticut River drainages.

It is a salamander that is slow to mature. Collections of Mudpuppies fall distinctly into size classes that represent age in years. The first six years are easy to group, with definite breaks in size. Because they do not lose their external gills, it is difficult to pinpoint when they transform from the larva to subadult stage. The larvae are distinctly patterned with two dark stripes running down their backs from head to tail. This striped pattern gradually disappears, and by about the fourth year, at 5.1 to 6.6 inches (13.0 to 16.7 cm), their color pattern looks much like a sexually mature adult. But they do not reach sexual maturity until they are about 7.9 inches (20.0 cm) long. Old adults have a blotchy mottled pattern. They are a long-lived salamander and may reach 30-plus years of age.

There are many reports of Mudpuppies migrating from deep lakes to shallow tributaries on an annual basis. Most nests have been found in water no more than a few feet deep, a depth that might be defined as a wading herpetologist's arm length. Deeper than that, and a researcher would have to fully submerge to check under rocks, a technique that probably didn't occur frequently before snorkeling and scuba became more common. Fisheries researchers have captured them using bottom trawls in water 60 feet (18 m) deep in Lake Erie, and fishermen have caught them ice fishing in depths of 90 feet (27 m) or more in Lake Champlain and the Finger Lakes. Whether they nest in waters that deep is not yet known, but they certainly use those deeper habitats to feed, as ice fishermen are catching them in midwinter on baited hooks.

There are two dominant threats to mudpuppies in the Northeast: lampricides and botulism. Sea Lampreys (*Petromyzon marinus*) scar and sometimes kill sport fish, primarily trout and salmon. Lampricides are used to kill the Sea Lampreys in the larger lakes and rivers of the Northeast. Although highly specific, lampricides do kill some "nontarget organisms," in particular, some amphibian larvae, small fishes such as darters and stonecats, and Mudpuppies. Mudpuppies at every life stage are very susceptible to lampricides. Immediately following lampricide treatment, Mudpuppies appear to be the only nontarget species that has been eliminated from the river systems.

A second threat to the Mudpuppy is Type E botulism. Die-offs from this cause are found mainly in Lake Erie and Lake Ontario on both the US and Canadian sides of the border. During a die-off in the late 1990s, one estimate found one dead Mudpuppy approximately every 10 feet (3 m) on a stretch of Lake Erie shoreline that extended from Buffalo, New York, to Erie, Pennsylvania—about 90 miles (145 km) of shoreline, or about 47,500 dead Mudpuppies. As appalling as the dead Mudpuppies were, it didn't stop there. The Mudpuppies were dying from feeding on Round Gobies, an introduced fish that was also a reservoir of the botulism. Gulls, diving ducks, and loons were feeding on the distressed and dying Mudpuppies as they floated ashore, and these birds were in turn dying in huge numbers.

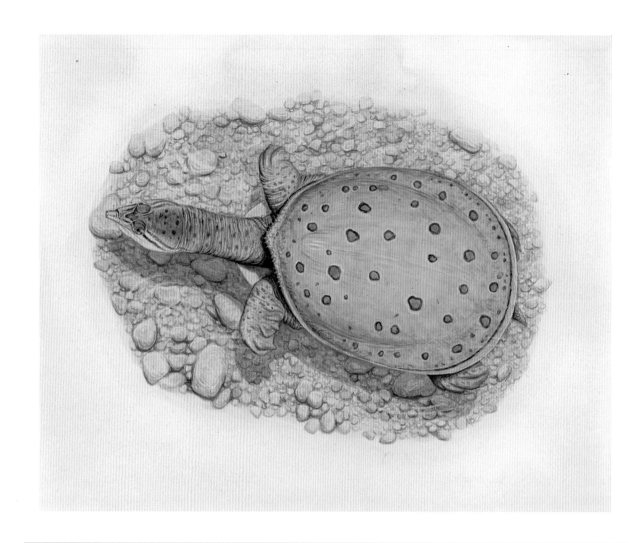

Spiny Softshell
Apalone spinifera

Type specimen described by Charles Alexandre Lesueur in 1827, collected from the
Wabash River near New Harmony, Indiana

Carapace length: hatchling = 1.3 inches (3.2 cm) to adult = 17.0 inches (43.2 cm)

Threatened in Vermont

The spiny part of the Spiny Softshell's name refers to the pointed, conical tubercles or bumps found behind the head on the edge of the carapace. The only other softshell turtle in the Northeast is the closely related Smooth Softshell (*Apalone mutica*), found in the Ohio River drainage of West Virginia and formerly at a few locations in western Pennsylvania. The Smooth Softshell lacks the pointed tubercles and is slightly smaller.

In a number of ways the Spiny Softshell is a very unturtle-like turtle. For one, it is a fast turtle, both on land and in the water. With its streamlined shape and fully webbed feet, it is not surprising that it is a strong swimmer. It is on land where its speed is astonishing. If startled or threatened on land, it doesn't awkwardly struggle to get back in water. It sprints! And because Spiny Softshells usually face the water when basking or nesting, they are poised to make that dash to the water! Spiny Softshells have been clocked on land at speeds of nearly 15 miles per hour (24 km per hour).

The second unturtle-like characteristic is the lack of a hard shell. As its common name implies, it has a soft shell rather than the rigid shell covered with horny scutes that is the standard feature of most turtles. Both the carapace and the plastron of the Spiny Softshell are formed by a flexible, leathery covering over the ribcage. While lacking the hard protective shell of other turtles, the Spiny Softshell makes up for it with sharp claws and strong mandibles, hidden by a fleshy covering, as its defensive weapons. Couple these features with a long neck, strong legs, and quick reflexes, and the Spiny Softshell can produce a painful bite or scratch.

The male and the female differ greatly in size, with the male reaching a maximum carapace length of just 9.3 inches (23.5 cm), or approximately half the length of the female. The surface of the male's carapace also has the texture of sandpaper, whereas the female's carapace feels smooth. The male's carapace is patterned with dark-bordered eye-like rings or spots. The female's carapace also has circular blotches when young but without the distinct ring-shaped marking. This pattern becomes more obscure as the female matures, and it almost disappears on really old females.

The Spiny Softshell approaches respiration in a slightly different way than most other turtles. Most typically it lies in shallow waters with just the tip of its nostrils at the end of its long tubular snout sticking above the surface of the water, allowing it to breathe while its body is hidden in the soft mud or sand of the lakeshore bottom or edge of a stream. It can also be seen basking as it floats just below the surface of the water with its signature snout projecting above the water. The Spiny Softshell thrives best in clean, well-oxygenated water where it can absorb dissolved oxygen directly from the water and expel carbon dioxide through membranes that line the mouth, pharynx, and cloaca.

Wood Turtle
Glyptemys insculpta

Type specimen described by John Eatton LeConte in 1830, collected from the
vicinity of New York City

Carapace length: hatchling = 1.1 inches (2.8 cm) to adult = 9.2 inches (23.4 cm)

Threatened in New Jersey and Virginia

Extirpated from the District of Columbia

The Wood Turtle is generally associated with rivers and streams with stretches of fast-flowing water alternating with deeper, quiet pools but is equally at home on land as in the water. The Wood Turtle is widespread in the region, being absent in portions of northern New York and reaching its southern limit in northern Virginia and the eastern panhandle of West Virginia.

The Wood Turtle gets its name from the sculpted carapace, which gives the impression of weathered cross sections of a tree. Of all northeastern turtle species, the Wood Turtle has the most prominent annuli, the concentric raised growth rings, on each of its scutes. These rings show distinct breaks between periods of rapid and slowed growth, with each ring corresponding to a stoppage of growth. Growth stops during the winter months when the Wood Turtle is in hibernation—more correctly known as brumation for cold-blooded animals—suggesting that the rings represent annual increments of growth. Other events also cause a slowing down or stoppage of growth, such as a serious injury or an unusually prolonged cold or hot spell during the active season. If the turtle slows down or stops feeding, a false annulus may be produced, so simply counting rings does not necessarily reflect the turtle's age. When the turtle reaches sexual maturity, at about 10 to 12 years of age, production of easily discernible annuli slows down, and it is rare to be able to count more than about 18 or 20 annuli.

In addition to the beautifully sculpted carapace with a slight middorsal keel, the Wood Turtle's neck and legs are decorated with a beautiful orange-red wash. The plastron is yellow with dark markings toward the edge of each plastral scute, giving the vague impression of a dried oak-leaf pattern.

The riverine habitat is key to the Wood Turtle's life history and survival. The turtle overwinters in the deeper pools, deep enough to avoid the buildup of ice, often partially covered with the dead leaves. Wood Turtles may also be found tucked up under tree roots along the banks or in the debris trapped by downed logs in the stream. In the spring, they may mate in the water or on land. They spend most of the summer in the floodplain alongside the water, but they may retreat to streams during extended periods of hot, dry weather.

Once they emerge from hibernation, Wood Turtles are found basking on the shore when air temperatures are greater than water temperatures. As the season warms, they remain on land for extended periods, feeding, searching for mates, and nesting. Most of this movement is parallel to the stream course, with turtles preferring the floodplain vegetation to the deeper forest cover. Males move farther than females, often journeying up and down stream corridors 2.0 miles (3.2 km) or more during their annual treks. In the Northeast, they often come in contact with roads, where crossing is a problem. Like many turtles, they seem to avoid traveling through culverts or even under many bridges that darken the stream channel. They are compelled to leave the water and cross the paved road surface, where they encounter motorized vehicles. The cars often win.

Up until the early 1900s, the Wood Turtle and the Box Turtle were sought as a food source. Unregulated collecting caused noticeable declines in Wood Turtle populations in the Northeast and led the State of New York in 1905 to pass perhaps the first law in the United States to protect any species of herpetofauna, forbidding the take of "land turtles and tortoises, including wood and box turtles." Although no longer pursued for food, even today the Wood Turtle is still the target of illegal collecting for the pet trade.

Northern Map Turtle
Graptemys geographica

Type specimen described by Charles Alexandre Lesueur in 1817, collected from a marsh on the borders of Lake Erie

Carapace length: hatchling = 1.2 inches (3.1 cm) to adult = 10.7 inches (27.3 cm)

Endangered in Maryland

Introduced in New Jersey

Known as the "Geographical Terrapin" in the 1800s and early 1900s, the Northern Map Turtle travels with a map on its back. Or so it seemed to early observers who noted the lines, polygons, and squiggles that decorated its carapace that were most obvious on younger individuals. The carapace has a low profile with a slight keel, or ridge, on its midline. The keel is more obvious on the male, appearing as a low series of bumps when viewed from the side. The female is significantly larger than the male, with males reaching a maximum size of 6.3 inches (15.9 cm).

This big-waters turtle is an excellent swimmer and frequent basker. It is easiest to observe where onshore and offshore rocks, logs, fallen trees, jetties, riprap, or other man-made platforms are numerous. Large groups of Map Turtles can often be seen congregating at one of these basking sites, sometimes so crowded that smaller turtles bask on the backs of larger turtles. In areas where their ranges overlap, Northern Map Turtles may be seen basking with Spiny Softshells, Red-eared Sliders, and Painted Turtles. Map Turtles travel 4 miles (6.4 km) or more up and down river from their overwintering sites for summer foraging or nesting. Their primary food is mollusks, but because the female is significantly larger than the male, she can feed on bigger and harder-shelled mollusks than the male. As an example of an environmental problem sometimes benefiting select populations of native species, Northern Map Turtles have taken advantage of the abundance of invasive mollusks such as Zebra Mussels (*Dreissena polymorpha*), Quagga Mussels (*D. rostriformis bugensis*), and Asian Clams (*Corbicula fluminea*).

The Northern Map Turtle is another example of a species that does not fall neatly into the description we all got in high school biology class of how vertebrates reproduce. Textbook reproduction involves genetic factors similar in humans and most other mammals, where the X and the Y chromosomes are sex chromosomes. A male has an X and a Y chromosome, and a female has two X chromosomes. Offspring have an equal possibility of being male or female. With Map Turtles and many other turtle species that lack X and Y chromosomes—including the common Snapping Turtle, the Red-eared Slider, the Painted Turtle, all sea turtles, all crocodilians, and some lizards—the temperature at which the eggs incubate determines the sex of offspring. This phenomenon is called temperature-dependent sex determination, or TSD, and can result in biased sex ratios. Depending on the incubation temperature, a nest can produce all males or all females.

Generally, for many turtles studied in the laboratory, cooler nest temperatures produce more males and warmer temperatures produce more females. For Map Turtles, the hatchlings will be primarily males if the nest incubates at 77°F (25°C), whereas at 86°F to 95°F (30°C to 35°C), the production of females will be favored. What happens in the wild? If the traditional nest site becomes shaded, are more males produced? If it is a particularly hot summer, will more females be produced? How does global warming affect sex ratio? These questions have not been adequately answered, nor has the question of whether there is an evolutionary advantage of TSD over genetic sex determination.

Northern Red-bellied Cooter
Pseudemys rubriventris

Type specimen described by John Eatton LeConte in 1830, collected from the
Delaware River near Trenton, New Jersey

Carapace length: hatchling = 1.0 inch (2.5 cm) to adult = 15.7 inches (40.0 cm)

Federally endangered (Massachusetts population only)

Endangered in Massachusetts

Threatened in Pennsylvania

Introduced to New York

The Northern Red-bellied Cooter is the largest basking turtle within its range, but it has the smallest total range of any turtle in the Northeast. Its common name suggests a bright red plastron, but it has more of a yellowish plastron with an orange or reddish wash. Females have vertical red lines on the costal scutes, whereas older males, slightly smaller than the females, have a generally red-brown carapace. In old individuals and in some populations, the carapace is nearly pure black.

This turtle is found from central New Jersey to North Carolina primarily east of Interstate 95, extending inland to the eastern panhandle of West Virginia through the Potomac River watershed. A disjunct population is found in eastern Massachusetts south of Boston. This population is nearly 200 miles (320 km) as the crow flies, and much farther as the turtle crawls, northeast of its northern range limit in New Jersey. The southern population is considered a river turtle found mostly in big flowing waters, like the Delaware, Susquehanna, Potomac, and Shenandoah Rivers. The northern population in Massachusetts is confined to a series of small, quiet-water ponds.

This separation, both geographically and by habitat, of these two population segments resulted in the northern population being considered a separate subspecies for many years. This separate subspecies, the Plymouth Red-bellied Turtle (*Pseudemys rubriventris bangsi*), became in 1980 the first freshwater turtle to be listed by the US Fish and Wildlife Service as an endangered species. Based on subsequent detailed genetic analysis, it is no longer considered a distinct subspecies, but the population is still commonly referred to as the Plymouth Red-belly, and it remains a distinct population of endangered turtle on the federal list.

When it was first listed, there were thought to be fewer than 300 geriatric Plymouth Red-bellied Turtles remaining, with little or no reproduction occurring, in part because of Bullfrog predation of hatchlings. Intensive efforts were begun to headstart turtles and relocate some of the headstarted turtles to unoccupied pond habitats. Headstarting can involve a number of activities. The first hurdle is to get through the egg stage. This can be accomplished by either protecting the nests in which females have placed eggs, by moving the nests to safer locations, or by transferring the eggs to a facility where they can be artificially incubated. The next stage is to release the hatchlings directly into the pond, eliminating the upland traverse where terrestrial predators might get them. Sometimes hatchlings are raised for a period of time in captivity so they can develop harder shells and a larger size, making them predator resistant if not actually predator proof. These headstarting efforts have paid off for the Plymouth Red-bellied Turtle. The population has increased about tenfold since recovery efforts first began.

Eastern Musk Turtle
Sternotherus odoratus

Type specimen described by Pierre André Latreille in 1801, collected from the
vicinity of Charleston, South Carolina

Total length: hatchling = 0.9 inches (2.2 cm) to adult = 5.4 inches (13.7 cm)

The Eastern Musk Turtle, also known as the Stinkpot, has a wonderfully descriptive name. When annoyed, it emits a vile odor from the four musk glands found under the carapace, behind the bridge. It also has a disproportionately large head for a turtle its size, equipped with a sharp beak and connected to a long neck. It can and will deliver a painful but not serious bite.

The Eastern Musk Turtle has a smooth, high-arched oval-shaped, olive-brown carapace with 11 marginal scutes, whereas most turtles have 12. Like the Snapping Turtle, the yellowish plastron is much reduced, offering little protection if flipped over by a predator. The tail of the male Stinkpot ends with a scale modified into a hard, pointed spine. On each side of its head are two yellow stripes and barbels under its chin and throat.

A highly aquatic turtle rarely seen far from water, Stinkpots are not accomplished swimmers. They are most frequently observed as they walk along the pond or rivershore in a few feet of water, especially around dusk. In such a situation they are bottom-feeding omnivores, eating both vegetation and small prey or scavenging larger food items that they can tear apart with sharp beaks and claws. They will take a baited hook, which often leads to a sad end. Whether repelled by the malodorous little turtle's stench, the pugnacious turtle's attempts to bite, the misidentification of a Stinkpot for a young Snapping Turtle, or the mistaken belief that it poses a threat to game fish, some anglers simply kill them rather than attempt to remove the hook. Fisheries biologists have reported capturing them in large numbers when bottom seining, especially in the fall or spring when Stinkpots congregate in deeper pools to overwinter.

Predators of this little turtle include mink, otters, and raccoons. Herons, Bullfrogs, bass, and watersnakes will take hatchlings and younger turtles. Nest predators such as raccoons and crows also are a threat. A pair of Bald Eagles had a nest at one small lake I visited. Numerous Painted Turtles could be seen basking around the shoreline. Upon checking the eagle nest, however, the five empty turtle shells found there were all Common Musk Turtles. The Musk Turtle's reduced plastron must have made it easier for the eagles to pick out the meat to feed their young. The eagles had clearly learned that it was more efficient to prey on the small turtles walking in shallow water than to pluck the basking Painted Turtles off their logs.

Many observers have noted the climbing ability of Eastern Musk Turtles. They bask on the branches or limbs of woody vegetation overhanging the edge of the river or lake up to 6 feet (1.8 m) above the water surface. If startled, they will simply drop into the water to escape. One observer quietly drifted his canoe under a basking Stinkpot that then dropped into his boat. I once observed a Stinkpot clinging to a sloping telephone pole guy-wire about 4 feet (1.2 m) above the water.

Curly Grass Fern

9 The Coastal Plain

I T IS only 730 air miles (1,175 km) from the point where Maine nestles against New Brunswick, Canada, to the point where Virginia touches North Carolina. Between these two points there is a combined distance of 13,468 miles (21,675 km) of shoreline. Large bays such as the Chesapeake and Delaware; estuaries; long, hook-shaped peninsulas like Cape Cod; major islands like Long Island; and numerous offshore and barrier islands contribute to the difference. The coastal plain also includes beaches, sand dunes, maritime forests, coastal plain wetlands, freshwater streams, and large areas of low-lying uplands that often extend many miles inland. From New Jersey south, the fall line separates the Piedmont from the coastal plain.

Estuaries extend many miles up some of the great rivers of the Northeast. The Hudson River, with a daily tidal flux of 3 to 4 feet (0.9 to 1.2 m) at the Troy Dam, 154 miles (248 km) from the coast, is the longest. This long tidal reach can carry brackish water many miles from the coast. In summer, it is not uncommon for the salt front to reach 60 miles (97 km) or more upstream. Salt in the water creates a huge challenge for amphibians and reptiles. Only the Sea Turtle and the Diamond-backed Terrapin thrive in a saltwater environment. Several other species tolerate the brackish waters including the Eastern Mud Turtle and Snapping Turtle.

Threats to coastal plain habitats include shoreline development, bulkheads, and maintenance of swimming beaches. Ocean level rise as a result of global climate change has already affected low-lying areas, and these impacts are expected to become more noticeable in the coming decades. Many miles of shoreline habitat have been lost in the vicinity of all the major coastal cities of the Northeast from Portland, Maine, to Norfolk, Virginia. But what is more amazing is to experience the habitats that remain.

Within the coastal plain there are numerous areas that experience heavy recreational use, such as Cape Cod in Massachusetts, Cape May in New Jersey, Acadia National Park in Maine, and ocean beaches on the Delmarva Peninsula and Long Island, New York. There are many national, state, and local parks that provide natural, relatively undisturbed, habitats for rare native plants and animals, such as the Northern Pinesnake, Eastern Tiger Salamander, Pine Barrens Treefrog, and Curly Grass Fern (*Schizaea pusilla*).

A large and diverse mix of freshwater, brackish, and saltwater habitats, the New Jersey Meadowlands is an area that has long been the subject of human encroachments. The area originally encompassed about 42 square miles (109 km²). When early Dutch settlers arrived, they considered the wetlands of the Meadowlands to be wastelands, clearing the Atlantic White Cedar (*Chamaecyparis thyoides*) forests and constructing dikes to drain the land so they could use it to grow crops and graze livestock. Today, about one-third of the original complex of wetlands—13 square miles (34 km²)—remains, and the Meadowlands is now better known as the home of the New York Giants and New York Jets.

Northern Pinesnake
Pituophis melanoleucus melanoleucus

Type specimen described by François Marie Daudin in 1803, collected from "Floride"

Total length: hatchling = 12.0 inches (30.5 cm) to adult = 83.0 inches (210.8 cm)

Threatened in New Jersey

Extirpated from Maryland

The Northern Pinesnake is a large, stout-bodied snake with irregularly shaped black blotches on a white to yellowish to light gray background color. This general color pattern is reflected in its specific name from the Greek, with *melanos* meaning "black" and *leukos* meaning "white." The genus name is also derived from the Greek, with *pitys* meaning "pine" and *ophios* meaning "serpent" or "snake." It is the longest snake in the Northeast. It is a surprisingly secretive species for such a large snake.

Pituophis melanoleucus is a widespread, patchily distributed species east of the Mississippi River consisting of three subspecies. Only the Northern Pinesnake is found in our region, where there are two widely disjunct populations. The most northern population is found in the New Jersey Pine Barrens. A southern population is found in the Blue Ridge Mountains and the Ridge and Valley physiographic region of western Virginia. The features these two areas have in common are the open pine woodlands and sandy substrate. They differ in that the pine barrens area is a low-elevation flatland interspersed with seasonal wetlands and streams. The Virginia habitat is located on high-elevation ridgetops and steep-sloped hillsides, with an understory of Mountain Laurel and Rhododendron added to the open pine woodlands.

The sandy soil is an important feature of the fossorial Northern Pinesnake's habitat because it enables the snake to create burrows in the loose substrate. The snake also takes advantage of underground tunnels created by small mammals or the hollows left by decaying stumps and roots of dead trees. Females burrow into the loose sand to make nests and deposits eggs. Both sexes use these burrows as overwintering sites, which may be used repeatedly over a number of years but not necessarily every year in succession. The Northern Pinesnake hunts for prey both underground and above the surface, taking many kinds of small mammals. The Northern Pinesnake is a powerful constrictor, enabling it to take rabbits, squirrels, and rats. It can simply grab smaller prey, such as mice and voles, with its mouth, swallowing them live without first subduing by constriction.

In the New Jersey Pine Barrens, a Northern Pinesnake population threatened by a proposed development was relocated to an area under conservation management. Part of the mitigation plan was to construct six artificial hibernacula for the Pinesnakes to use for overwintering. The hibernacula were successfully used by both relocated and native snakes. For unknown reasons, however, long-term survival of the relocated snakes was minimal.

Eastern Tiger Salamander
Ambystoma tigrinum

Type specimen described by Jacob Green in 1825, collected from Moorestown, New Jersey

Total Length: metamorph = 3.5 inches (8.9 cm) to adult = 13.0 inches (33.0 cm)

Endangered in Delaware, Maryland, New Jersey, New York, Virginia

Extirpated from Pennsylvania

The Eastern Tiger Salamander is the largest terrestrial salamander in the Northeast. It is basically a dull brown to greenish-black stocky salamander with irregular, elongate light yellow to yellow-olive spots on its back and sides with a yellowish belly. It may take five years to reach sexual maturity and can live up to 20 years.

The Eastern Tiger Salamander breeds in seasonal wetlands, doing best in pools where there are no fish. Its migration to breeding ponds may begin in late December or not until after the first of April, depending on latitude and the severity of the winter. Females migrate several days after the males have entered the pond. A cold snap between the two migrations may result in the males patiently waiting under the newly formed ice for several days or even weeks for the females to arrive. Egg laying occurs from mid-January until early April. The newly hatched larvae, measuring barely 0.5 inches (1.3 cm) in length, grow quickly, reaching 4.0 inches (10.2 cm) by early summer. The larvae develop extremely bushy external gills typical of salamanders that breed in ponds where dissolved oxygen in the water column is at a premium. Depending on conditions, the larvae are ready to transform and move out on land within 2.5 to 5 months.

A conflict arises when Tiger Salamanders occupy seasonal pools in areas where people view the pools as simply mosquito-breeding ponds. Various methods for controlling mosquitoes have been tried, some more harmful to the salamanders than others. Draining of the wetlands was an early solution. Chemical and biological controls have also been tried. One biological method is using Bti (*Bacillus thuringienis israelensis*), a highly selective bacterium that kills mosquito larvae without directly harming fish, birds, or mammals. Another method currently being promoted is the introduction of mosquitofish (*Gambusia* spp.), nonnative fish species that prey on mosquito larvae. Mosquitofish also consume tiger salamander larvae until the salamander larvae are big enough to reverse the predator–prey relationship. Tiger salamander larvae also consume mosquito larvae, so perhaps the best approach is to encourage a healthy population of ambystomid salamanders.

The Eastern Tiger Salamander was previously considered to be one of six subspecies of tiger salamanders but is now considered a separate species. It is widespread, with its population center in the upper Midwest where they occur in great numbers. In the East, however, Eastern Tiger Salamanders are pretty much confined to coastal areas from Long Island to Virginia and are rare and declining throughout their northeastern range. Intensive surveys, conducted in all these states, indicate that there are more Eastern Tiger Salamander populations on Long Island than all the other northeastern states combined. If the stronghold in the Northeast is in a rapidly developing area like Long Island, where land values are high, its future in our region cannot be considered secure.

Many-lined Salamander
Stereochilus marginatus

Type specimen described by Edward Hallowell in 1856, collected from
Liberty County, Georgia

Total length: metamorph = 2.2 inches (5.5 cm) to adult = 4.5 inches (11.5 cm)

The Many-lined Salamander is a rather plain-looking brown to dull yellow salamander with many faint dark and light alternating parallel lines or streaks on its sides. In some individuals these lines are indistinct dashes or a series of dots. Its yellow belly is marked with scattered dark specks. The Many-lined Salamander has a small, flattened, wedge-shaped head and a short, keeled tail equal in length to a bit less than half the salamander's total length, providing a streamlined shape for an aquatic environment.

This is a highly aquatic species that inhabits woodland ponds, tupelo (*Nyssa* spp.) and Bald Cypress (*Taxodium distichum*) swamps, sluggish streams, and other permanent aquatic habitats, including man-made ditches, borrow pits, and canals. It can be found from the extreme southeastern coastal plain region of Virginia to the northeasternmost region of Florida. It does leave the water but will stay close to the margin of its aquatic home, hiding under logs, wet leaf litter, tussocks of *Sphagnum* moss, or other wet microhabitats.

The Many-lined Salamander is a fall breeder. Females usually nest in water but will nest on land close enough to the stream or pond so that hatchlings have access to an aquatic habitat after they emerge from the eggs. The females frequently remain with the nest to brood the eggs if placed on land. In slow-moving streams or ponds, the eggs may be attached to water moss (*Fontinalis* spp.), which grows in long, stringy branches attached to rocks. The larvae when first hatching are tiny, only about a third of an inch (0.8 cm) long, and are distinctly bicolored with a dark brown dorsum and yellow ventral surface. Unlike most pond-dwelling salamanders, the larvae have functional legs at hatching. The larvae grow slowly, taking from 1.5 to 3 years before transforming into metamorphs.

The main food items for both larvae and adults are aquatic invertebrates, differing only in size of the prey. Most of their diet is composed of small crustaceans such as amphipods (scuds and side swimmers), isopods (aquatic sow bugs), and ostracods (shelled organisms that superficially resemble tiny clams), as well as aquatic insect larvae such as chironomids (nonbiting midges). In turn, dragonfly and dytiscid beetle (predaceous diving beetles) larvae, aquatic snakes, fish, and wading birds prey upon Many-lined Salamanders.

Pine Barrens Treefrog
Hyla andersonii

Type specimen described by Spencer Fullerton Baird in 1854, reportedly from the vicinity of Anderson, South Carolina

Head–body length: metamorph = 1.0 inch (2.5 cm) to adult = 2.0 inches (5.1 cm)

Endangered in New Jersey

The Pine Barrens Treefrog is instantly recognized by the purple stripe along its sides bordered on the upper edge by a thin yellow or white line, and a plain bright green back. Its belly is white. When sitting quietly, the exposed surfaces of both front and rear legs match the green of its back, while the hidden surfaces of the legs are a bright orange with yellow dots similar to the Gray Treefrog's legs.

It is one of the most attractive frogs in the Northeast, but it is also the rarest. The Pine Barrens Treefrog is found only in the New Jersey Pine Barrens area within the Northeast. Two other widely disjunct populations of this species occur; the closest one is about 340 miles (550 km) south, extending from the Sandhills of North Carolina through the coastal plain into South Carolina. A third population occurs another 360 miles (580 km) southwest, in the western panhandle of Florida and adjacent Alabama. The original type locality identified by S. F. Baird in Anderson, South Carolina, was well outside of its current and known historic range. Karl P. Schmidt tried to correct that error in 1953 by revising the type locality to Aiken County, South Carolina, a location also outside of known range but closer to Richmond County, Georgia, where Wilfred T. Neill reported it in 1947. Neill's location is also outside of the Pine Barrens Treefrog's known range. Perhaps we will never know the location of the original type, but we may find there are as yet undiscovered populations of this treefrog. After all, three widely spaced dots on a map had to have been connected at some time in the past.

The quest to determine where this species was first found has been elusive. But even within its known range, the Pine Barrens Treefrog is an elusive species. It is found in the more acidic habitats such as bogs, Atlantic White Cedar swamps, and pocosins surrounded by pine forest. The Pine Barrens Treefrog is not one that is frequently found by stumbling across it on a random tramp through the woods, a method that does work for Spring Peepers, Wood Frogs, and Gray Treefrogs. It is rarely found when searching for it outside of the breeding season. It is best to wait until breeding season and follow the sound of its advertisement call, a series of high-pitched and melodic but somewhat nasal *quonk-quonk-quonk-quonks* that males use to attract females. Once one male begins, other males join the chorus. Focusing on the call, the treefrog could be perched on the ground, low in a tree, or at an elevation in between from a shrub. As Alexander B. Klots pointed out in 1930, without the call to focus on, "A silent *andersonii* in a thick tangle of Blueberry bushes could give points to a very small needle in a very large haystack."

Green Treefrog
Hyla cinerea

Type specimen described by Johann Gottlob Schneider in 1799, collected from the vicinity of Charleston, South Carolina

Head–body length: metamorph = 0.5 inches (1.2 cm) to adult = 2.5 inches (6.4 cm)

Introduced to West Virginia

The Green Treefrog is one of the larger treefrogs in the Northeast and, within its range in the coastal plain from the Delmarva Peninsula south, it is one of the most common treefrogs. Its vivid bright green to lime-green color and sharply defined white stripe along each side give it a kind of regal appearance and make it easy to identify. The white line may be bordered in black or in some individuals missing altogether. Its back may be marked with small white to gold dots outlined with a thin black border. Its body is longer and more slender than that of the Pine Barrens Treefrog.

If you suspect the frog you are observing is a Green Treefrog but it does not precisely match this description, check the temperature, background where it is sitting, and the season before you rule out Green Treefrog. An individual Green Treefrog's color is dependent on all these factors. The lighter and brighter greens are evident at higher temperatures. Frogs exposed to temperatures of 80.0°F (26.7°C) exhibit the brightest colors. Frogs held at temperatures of 50.0°F (10.0°C) are dull green to slate gray. During breeding season from May through early July, the overall color of the male may be more yellow than green. During cooler periods, or when the frog has retreated to a microhabitat where the temperature is significantly lower, the Green Treefrog will be a duller green to gray color. A similar response occurs when frogs are placed on substrates with different levels of brightness. A bright background produces a bright treefrog, and a dark or dull background produces a dull-colored treefrog. The Green Treefrog's strongest choruses occur when the air temperature is 65.0°F (18.3°C) or warmer.

The advertisement call of the Green Treefrog sounds to some observers like a bell, giving it the colloquial names of Bell Frog or Cowbell Frog. The choruses have a nasal or tinny quality. Choruses of hundreds, if not thousands, of frogs calling in concert from the shores of ponds, swamps, or riverbanks can be heard at considerable distances. The Green Treefrog is more fish tolerant than other members of its genus and will breed in permanent ponds, streams, and wetlands. It does, however, prefer ponds with abundant floating and emergent vegetation. Females lay numerous small clumps of eggs from early May to late July. The eggs hatch within a week, and the tadpoles transform into subadults in about two months.

Southern Leopard Frog
Lithobates sphenocephalus

Type specimen described by Edward Drinker Cope in 1886, collected near the
St. John's River, Florida

Head–body length: metamorph = 0.7 inches (1.8 cm) to adult = 5.0 inches (12.7 cm)

Endangered in Pennsylvania

The Southern Leopard Frog looks similar to the Northern Leopard Frog and was for years considered simply as a subspecies of *Lithobates pipiens*. Older naturalists might know the Southern Leopard Frog as *Rana utricularia*. *Utricularia* is from the Latin meaning "player-on-the-bagpipes," a reference to the two large vocal sacs sported by this species, one on each side of its throat. If you knew where you were in the Northeast, a quick check of the range map would reveal which leopard frog was in your area because they have nonoverlapping ranges, a trait known as being allopatric in biologist terminology. If not satisfied with an identification based on a range map, check the tympanum. The Southern Leopard Frog will have a white dot near the center of the tympanum. On the Northern Leopard Frog the spot will be light tan or missing all together. The voice, which is a guttural clucking or chortling sound in both, will also separate them. The Southern Leopard Frog's call is slower, only about 13 pulses per second, to the Northern Leopard Frog's rapid 20 pulses per second.

Recognition of the Southern Leopard Frog as a distinct species did not end the confusion concerning leopard frogs in the Northeast. Perhaps this should have been expected. Researchers, led by Jeremy Feinberg, have recently (circa 2012) noticed that the call of the leopard frogs they were hearing did not fit cleanly as either Northern or Southern Leopard Frog. And where did they find this "new" frog? On Staten Island, part of New York City, the city with the largest human population in the United States! Basically this new frog was hidden in plain sight until its call, followed by careful genetic analysis, revealed a third species of leopard frog in the Northeast, the Atlantic Coastal Leopard Frog, named *L. kauffeldii* in honor of Carl Kauffeld, longtime curator of the Staten Island Zoo. The species is now known from central Connecticut south to North Carolina, but there certainly is more to learn about its habitat preferences, distribution, and life history.

It is too soon to determine to what extent the ranges of the Southern Leopard Frog and the Atlantic Coastal Leopard Frog overlap. Maybe the true Southern Leopard Frog never did quite get as far north as New York and Connecticut. Are there more cryptic species of leopard frog out there that we have simply overlooked because we didn't have the proper tools to identify them? Will our understanding of these fascinating creatures continue to change over the next decade or two? My guess is yes, they will. These are the kinds of questions that excite taxonomists, geneticists, and evolutionary biologists and that challenge biologists who are working on the conservation issues facing these species.

Carpenter Frog
Lithobates virgatipes

Type specimen described by Edward Drinker Cope in 1891, reported from a cutoff of a tributary of the Great Egg Harbor River, Atlantic County, New Jersey

Head–body length: metamorph = 0.9 inches (2.3 cm) to adult = 2.6 inches (6.7 cm)

The call of the Carpenter Frog sounds like a carpenter driving a nail into a board, with an echo following each strike in rapid succession. As a chorus of many frogs, the sound becomes that of a room full of young kids with hammers pounding on wooden benches. A chorus is nothing without an audience. The Carpenter Frog is a good species for a discussion of the variety of calls a frog makes and what other frogs are hearing.

Begin by taking note of the size of the "ear," or tympanum, the round patch of skin just behind the eye that is more commonly called the eardrum. The tympanum in frogs, as in other vertebrates, serves as the receiver of sound waves. Once detected, these sound waves are transported to the middle and inner ear and then to the brain, which interprets the signal. The size of the tympanum in Carpenter Frogs and five other northeastern ranid frogs can be used to determine sex. In males, the tympanum is significantly bigger than the eye. In females, it is about the same size as the eye or slightly smaller. The sound received and interpreted is a function of the size of the tympanic membrane.

Frogs use their vocal sacs to project several types of calls. The Carpenter Frog, like leopard frogs, has two vocal sacs, one on each side of its head, rather than a single vocal sac like the Bullfrog or Green Frog. The typical chorus we hear are the males calling, not the females. The mating or advertisement call is used to attract females and at the same time let other males know that this spot is occupied and that the calling frog is ready to defend his territory, so "back off!" The males respond by spacing themselves a comfortable distance from their closest neighbors. If a neighboring frog does get too close, an aggressive or territorial call warns the intruding frog to back away to avoid a fight. Frogs also make a dramatic and sudden distress call when being attacked by a predator, and a warning call when they are startled to tell other frogs to be careful, that something dangerous may be afoot. Territorial and mating calls are usually only made by male frogs searching for a mate. The other calls are made by both male and female frogs.

We use these calls, mostly the advertisement call, to identify the species of frog that made the call. But the frogs themselves hear things differently. The male with the bigger tympanum hears the lower-frequency parts of the call, and the female with the smaller tympanum hears the higher frequencies. Same call, different messages.

Eastern Mud Turtle
Kinosternon subrubrum subrubrum

Type specimen described by Bernard-Germain-Etienne de le Ville-sur-Illon Lacèpede
in 1788, collected from the vicinity of Philadelphia, Pennsylvania

Carapace length: hatchling = 0.7 inches (1.7 cm) to adult = 4.9 inches (12.4 cm)

Endangered in New York

The Eastern Mud Turtle is a nondescript, plain, dark olive drab to brown or almost black, elliptically rounded turtle that is similar in shape to the Stinkpot but slightly smaller. The Eastern Mud Turtle does, however, have an attractive yellow-brown plastron with a double hinge, enabling it to partially close both the front and back lobes of its lower shell. Like the Stinkpot, it has just 11 marginal scutes on each side of its carapace. The lower shells of the hatchlings have a distinct reddish hue with a dark center; hence the species name *subrubrum*, meaning "below red," or more literally "red plastron." In the southeastern coastal plain region of Virginia, there is a closely related species, the Striped Mud Turtle (*Kinosternon baurii*), which has stripes on both its carapace and the sides of its head.

In the Northeast, the Eastern Mud Turtle's range extends from brackish coastal waters as far north as Long Island, New York, to the interior Piedmont region of Virginia. It is one of the few freshwater turtles that can tolerate brackish water habitats. Those low-elevation coastal habitats in the North are threatened by sea level rise due to global climate change. Except for some of the Sea Turtles that seasonally range as far north as New England, the Eastern Mud Turtle is the rarest turtle in New York. Farther south, they are locally abundant in shallow ponds and swamps in southeastern Virginia.

Eastern Mud Turtles are equally at home on land and in the water. But, like Stinkpots, they are not good swimmers. In the water they are more likely to be seen walking slowly along the bottom of a pond or slow-moving stream. Some individuals overwinter buried in the soft mud at the bottom of the pond. Others may spend most of the summer on land and dig their own burrows to overwinter in sandy soil as much as a foot below the surface.

They never reach a size to become predator proof. Foxes, raccoons, coyotes, and Bald Eagles can easily kill a fully grown adult Eastern Mud Turtle. Predation, along with development in coastal areas, has caused significant declines in some populations. The young are basically carnivorous, but adult Eastern Mud Turtles are omnivores and scavengers, foraging on such food items as the seeds of White Water Lilies (*Nymphaea odorata*), crayfish, mollusks, amphibians, and carrion.

Northern Diamond-backed Terrapin
Malaclemys terrapin terrapin

Type specimen described by Johann David Schoepff in 1793, collected from the coastal waters of Long Island, New York

Carapace length: hatchling = 1.0 inch (2.5 cm) to adult = 9.0 inches (22.9 cm)

Threatened in Massachusetts

State reptile of Maryland and official mascot of the University of Maryland

Diamond-backed Terrapins are strictly saltwater or brackish-water turtles found in coastal waters and estuaries from Massachusetts to Texas. Seven subspecies have been named, but only the Northern Diamond-backed Terrapin is found in the Northeast, ranging from Cape Cod, Massachusetts, to Cape Hatteras, North Carolina, with the Chesapeake Bay hosting the largest concentration of this subspecies. It is about as close as a turtle can come to being a Sea Turtle without actually being a Sea Turtle. Its extensively webbed feet make it an adept swimmer; however, it does not make the long migrations to nest like Sea Turtles do. It winters at the bottom of estuaries and nests on nearby beaches or dunes.

The Northern Diamond-backed Terrapin has a carapace attractively marked with concentric rings or annuli, a gray head and legs decorated with black spots, and white "lips." In quiet water, the terrapin can frequently be seen basking offshore as it floats with just its head above water. The white lips, its mandibles, are easily observed with binoculars or spotting scope. The female is decidedly larger than the male, which reaches a maximum carapace length of about 5.5 inches (14.0 cm); males are only about one-quarter the weight of females.

This turtle has been the most heavily exploited turtle since colonial times. The early Dutch historian Adriaen Van der Donck noted in 1656 that "some persons prepare delicious dishes from the water terrapin, which is luscious food." Commercial harvest for food in the late 1800s, which reached 400,000 pounds (181,437 kg) of turtles per year, led to population declines that were obvious before the turn of the century. Terrapin flesh has always been highly prized, so as the species declined, the price increased. In the early 1900s, an attempt to farm-raise terrapins for sale to restaurant markets began at Beaufort, North Carolina. This effort proved unprofitable and was abandoned in the 1930s. Following the Second World War, interest in Diamond-backed Terrapins for food increased, especially in Asian communities. By the 1990s, most states had passed regulations to prohibit unrestricted harvest of terrapins. Maryland adopted a complete prohibition on the collection of terrapins in 2007. Limited harvesting of terrapins for commercial or personal use is still allowed in Delaware, New Jersey, and New York. Several states are now requiring that a turtle excluder device be placed on all crab pots to prevent terrapins from entering a trap and drowning.

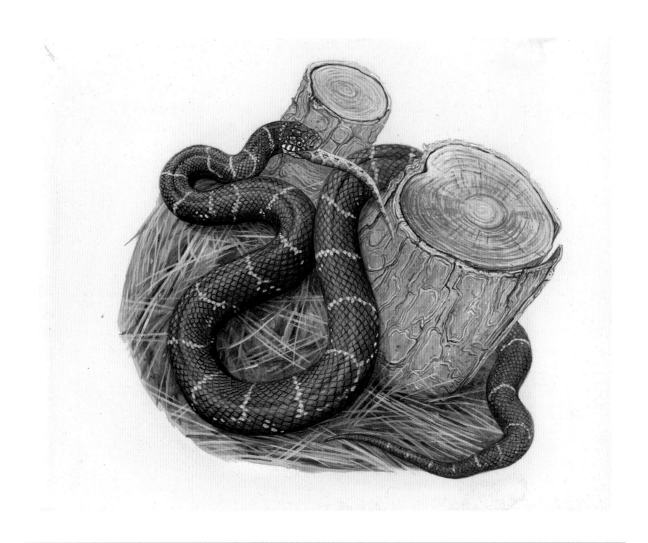

Eastern Kingsnake
Lampropeltis getula

Type specimen described by Carolus Linnaeus in 1766, collected from the vicinity of Charleston, South Carolina

Total Length: hatchling = 9.1 inches (23.0 cm) to adult = 82.0 inches (208.3 cm)

The Eastern Kingsnake is a large, glossy, smooth-scaled constrictor found from the New Jersey Pine Barrens south through the coastal plain, extending into the Piedmont in Virginia. The snake is marked with large black blotches separated by thin white bold lines, giving it a chain-like appearance and one of its other common names, the Chain Snake.

Its demeanor when startled or threatened is to rear up and hiss, vibrate its tail, and strike. It will bite if picked up but quickly settles down. It has been a favorite pet since the early 1900s among herpetoculturists who like to possess large, but not necessarily huge or dangerous, snakes. It readily takes to captivity and is easily handled. In captivity as well as in the wild, they readily eat a variety of small mammals. They should not be housed with other snakes, because the Eastern Kingsnake is well known as an ophiophage, or snake eater. It was presumably given the name "king" for just that reason, for a snake that eats other snakes is a king in many people's eyes. With its reputation as a snake eater, the Eastern Kingsnake has another advantage in that it does not even vaguely look like any of our venomous species, so it is not often killed because of mistaken identity. Furthermore, it is well known for its ability to consume venomous Copperheads, rattlesnakes, and Cottonmouths. Being immune to the venom of pit vipers, the Eastern Kingsnake wins most battles.

The Eastern Kingsnake is chiefly a terrestrial species, but it is often found along the edge of ponds, stream banks, or swamps in search of prey such as watersnakes or Rainbow Snakes. It uses its head to dig up favorite foods such as turtle eggs or to root under logs or other cover objects for other snakes or mice. The list of non-venomous snakes that Eastern Kingsnakes are known to consume includes Black Racers, Eastern Ratsnakes, gartersnakes, Ring-necked Snakes, earthsnakes, wormsnakes, Red-bellied Snakes, and Hog-nosed Snakes—essentially every species of snake they encounter. Although they are primarily terrestrial and will readily take to water, Eastern Kingsnakes are not climbers and are rarely encountered above the ground even as high as a low shrub. They will take ground-nesting bird eggs, birds if they can catch them, skinks, and other lizards.

A snake with the reputation of the Eastern Kingsnake can be expected to have a variety of colloquial names and myths. One such name is Thunder Snake, and the belief that if one is killed a thunderstorm will follow. As with the Kingsnake itself, a thunderstorm could be something to fear or something to welcome, depending where on the food chain you sit.

Appendixes

Appendix A
Habitats Utilized by Species of Frogs

	Northeastern Deciduous Forest	Dry Pine Woodlands	Northeastern Grasslands	Wicked Big Puddles	Bogs	Headwaters	Small Waters	Big Waters	The Coastal Plain
Spadefoot Toads: Pelobatids									
Eastern Spadefoot		■	■	■					■
True Toads: Bufonids									
American Toad	■	■	■	■	■	■	■		■
Fowler's Toad	■	■	■	■	■		■		■
Treefrogs: Hylids									
Eastern Cricket Frog	■		■	■	■	■			■
Pine Barrens Treefrog		■	■		■		■		■
Green Treefrog	■	■		■			■		■
Gray Treefrog	■	■	■	■			■		■
Spring Peeper	■	■	■	■	■		■		■
Western Chorus Frog	■	■			■				
Upland Chorus Frog	■	■	■	■	■		■		■
True Frogs: Ranids									
American Bullfrog			■			■	■	■	■
Carpenter Frog				■	■		■		■
Green Frog	■		■	■	■	■	■	■	■
Mink Frog	■				■	■	■	■	
Wood Frog	■			■					■
Northern Leopard Frog	■		■	■	■	■	■	■	
Southern Leopard Frog	■	■	■	■	■				■
Pickerel Frog	■		■		■	■	■	■	■

Appendix B
Habitats Utilized by Species of Salamanders

	Northeastern Deciduous Forest	Dry Pine Woodlands	Northeastern Grasslands	Wicked Big Puddles	Bogs	Headwaters	Small Waters	Big Waters	The Coastal Plain
Giant Salamanders: Cryptobranchids									
Eastern Hellbender						■		■	
Mudpuppies and Waterdogs: Proteids									
Common Mudpuppy						■	■	■	
Mole Salamanders: Ambystomids									
Marbled Salamander	■	■		■					■
Jefferson Salamander	■	■	■	■		■	■		
Blue-spotted Salamander	■	■	■	■		■	■		■
Spotted Salamander	■	■	■	■	■	■			■
Eastern Tiger Salamander	■	■	■	■			■		■
Newts: Salamandrids									
Red-spotted Newt	■	■	■	■					
Lungless Salamanders: Plethodontids									
Northern Dusky Salamander	■					■	■	■	
Allegheny Mountain Dusky Salamander	■					■			
Eastern Red-backed Salamander	■	■							■
Wehrle's Salamander	■	■							
Northern Slimy Salamander	■	■							
Yonahlossee Salamander	■								
Four-toed Salamander	■	■		■	■	■	■		■
Many-lined Salamander						■	■		■
Green Salamander	■	■		■		■			
Spring Salamander	■	■				■			
Mud Salamander	■	■			■		■		■
Northern Red Salamander	■			■	■	■	■		■
Northern Two-lined Salamander	■	■		■	■	■		■	
Eastern Long-tailed Salamander	■	■		■		■	■		

Appendix C
Habitats Utilized by Species of Turtles

	Northeastern Deciduous Forest	Dry Pine Woodlands	Northeastern Grasslands	Wicked Big Puddles	Bogs	Headwaters	Small Waters	Big Waters	The Coastal Plain
Snapping Turtles: Chelydrids									
Snapping Turtle	■		■	■	■		■	■	■
Mud and Musk Turtles: Kinosternids									
Eastern Musk Turtle				■			■	■	■
Eastern Mud Turtle		■	■	■	■				■
Pond and Marsh Turtles: Emydids									
Spotted Turtle	■			■	■		■		■
Bog Turtle	■				■		■		
Wood Turtle	■		■		■	■	■	■	
Eastern Box Turtle	■	■	■	■	■		■	■	■
Northern Diamond-backed Terrapin									■
Northern Map Turtle			■					■	
Yellow-bellied Slider				■			■	■	■
Northern Red-bellied Cooter							■	■	■
Painted Turtle	■		■	■	■		■	■	■
Blanding's Turtle	■		■	■	■		■	■	
Softshell Turtles: Trionychids									
Eastern Spiny Softshell							■	■	

Appendix D
Habitats Utilized by Species of Lizards

	Northeastern Deciduous Forest	Dry Pine Woodlands	Northeastern Grasslands	Wicked Big Puddles	Bogs	Headwaters	Small Waters	Big Waters	The Coastal Plain
Spiny Lizards: Phrynosomatids									
Eastern Fence Lizard	■	■	■						■
Skinks: Scincids									
Little Brown Skink	■	■	■				■		■
Common Five-lined Skink	■	■	■						■
Broad-headed Skink	■	■	■				■		■
Northern Coal Skink	■	■	■			■	■		

Appendix E
Habitats Utilized by Species of Snakes

	Northeastern Deciduous Forest	Dry Pine Woodlands	Northeastern Grasslands	Wicked Big Puddles	Bogs	Headwaters	Small Waters	Big Waters	The Coastal Plain
Nonvenomous Snakes: Colubrids									
Northern Watersnake			■	■	■	■	■	■	■
Queensnake						■		■	
Northern Brownsnake	■	■	■		■				■
Northern Red-bellied Snake	■	■	■		■				■
Common Gartersnake	■	■	■	■	■	■	■	■	■
Short-headed Gartersnake	■								
Eastern Ribbonsnake	■	■		■		■	■		■
Smooth Earthsnake	■		■						■
Eastern Hog-nosed Snake	■	■	■		■				
Ring-necked Snake	■	■	■		■				■
Eastern Wormsnake	■	■							■
Common Rainbow Snake				■		■	■		
Northern Black Racer	■	■	■	■	■				
Northern Rough Greensnake	■	■	■	■	■	■		■	■
Smooth Greensnake	■	■	■				■		■
Red Cornsnake	■	■							■
Eastern Ratsnake	■	■	■						■
Northern Pinesnake	■	■							■
Eastern Kingsnake	■	■	■			■	■		■
Eastern Milksnake	■	■	■						■
Northern Scarletsnake	■	■	■						■
Pit Vipers: Crotalids									
Northern Copperhead	■	■							■
Eastern Massasauga	■		■		■		■		
Timber Rattlesnake	■	■	■		■				■

Suggested Reading

For information about the major amphibian and reptile organizations in the United States, see:

American Society of Ichthyologists and Herpetologists: http://www.asih.org/
Center for North American Herpetology: http://www.cnah.org/
Herpetologists' League: http://www.herpetologistsleague.org/
Northeast Partners in Amphibian and Reptile Conservation: http://northeastparc.org/
Partners in Amphibian and Reptile Conservation: http://www.parcplace.org/
Society for the Study of Amphibians and Reptiles: https://ssarherps.org/

For more information on laws and regulations pertaining to the amphibians and reptiles in your state, see the website of the Association of Fish and Wildlife Agencies, which has links to each state's natural resource agency: http://www.fishwildlife.org/index.php?section=social-media

Behler, John L., and F. Wayne King. 1997. *National Audubon Society Field Guide to North American Amphibians and Reptiles.* Alfred A. Knopf, New York, NY.

Braun, E. Lucy. 1950. *Deciduous Forests of Eastern North America.* Blakiston, Philadelphia, PA.

Crother, Brian I. 2012. *Scientific and Standard English Names of Amphibians and Reptiles of North America North of Mexico, with Comments Regarding Confidence in Our Understanding*, 7th ed. Herpetological Circular No. 39. Society for the Study of Amphibians and Reptiles, Salt Lake City, UT.

Elliot, Lang, Carl Gerhardt, and Carlos Davidson. 2009. *The Frogs and Toads of North America: A Comprehensive Guide to Their Identification, Behavior, and Calls* [including CD of frog calls]. Houghton Mifflin Harcourt, New York, NY.

Ernst, Carl H., and Jeffrey E. Lovich. 2009. *Turtles of the United States and Canada*, 2nd ed. Johns Hopkins University Press, Baltimore, MD.

Gibbs, J. P., A. R. Breisch, P. K. Ducey, G. Johnson, J. L. Behler, and R. C. Bothner. 2007. *The Amphibians and Reptiles of New York State: Identification, Life History, and Conservation.* Oxford University Press, New York, NY.

Green, N. Bayard, and Thomas K. Pauley. 1987. *Amphibians and Reptiles in West Virginia.* University of Pittsburgh Press, Pittsburgh, PA.

Harding, James H. 1997. *Amphibians and Reptiles of the Great Lakes Region*. University of Michigan Press, Ann Arbor, MI.

Hulse, Arthur C., C. J. McCoy, and Ellen Censky. 2001. *Amphibians and Reptiles of Pennsylvania and the Northeast*. Comstock, Ithaca, NY.

Hunter, Malcolm L. Jr., Aram J. K. Calhoun, and Mark McCollough. 1999. *Maine Amphibians and Reptiles* [including CD of frog calls]. University of Maine Press, Orono, ME.

Klemens, Michael W. 1993. *Amphibians and Reptiles of Connecticut and Adjacent Regions*. Connecticut Geological and Natural History Survey Bulletin No. 112. Connecticut Department of Environmental Protection, Hartford, CT.

Mitchell, Joseph C. 1994. *The Reptiles of Virginia*. Smithsonian Institution Press, Washington, DC.

Mitchell, Joseph C., Alvin R. Breisch, and Kurt A. Buhlmann. 2006. *Habitat Management Guidelines for Amphibians and Reptiles of the Northeastern United States*. Partners in Amphibian and Reptile Conservation Technical Publication HMG-3, Montgomery, AL.

Petranka, James W. 1998. *Salamanders of the United States and Canada*. Smithsonian Institution Press, Washington, DC.

Pfingsten, Ralph A., Jeffrey G. Davis, Timothy O. Matson, Gregory J. Lipps Jr., Doug Wynn, and Brian J. Armitage (Eds.). 2013. *Amphibians of Ohio*. Ohio Biological Survey, Columbus, OH.

Powell, Robert, Roger Conant, and Joseph T. Collins. 2016. *Peterson Field Guide to the Reptiles and Amphibians of Eastern and Central North America*, 4th ed. Houghton Mifflin, New York, NY.

Schwarz, Vicki, and David M. Golden. 2002. *Field Guide to Reptiles and Amphibians of New Jersey*. New Jersey Division of Fish and Wildlife, Woodbine, NJ.

Smith, Hobart M. 1982. *Amphibians of North America: A Guide to Field Identification*. Golden Press, New York, NY.

Smith, Hobart M., and Edmund D. Brodie Jr. 1982. *Reptiles of North America: A Guide to Field Identification*. Golden Press, New York, NY.

Taylor, James. 1993. *The Amphibians and Reptiles of New Hampshire*. New Hampshire Fish and Game Department, Concord, NH.

White, James F. Jr., and Amy W. White. 2007. *Amphibians and Reptiles of Delmarva*. Tidewater Publishers, Centreville, MD.

Index

Boldface page numbers indicate illustrations